Building a Climate Resilient Economy and Society

'Earth provides enough to satisfy every man's needs,
but not every man's greed.'
Mahatma Gandhi

Building a Climate Resilient Economy and Society

Challenges and Opportunities

Edited by

K.N. Ninan

Chairperson, Centre for Economics, Environment and Society, Bangalore, India and Co-Chair, Panel on Methodological Assessment of Scenarios and Models of Biodiversity and Ecosystem Services, Intergovernmental Platform on Biodiversity and Ecosystem Services, United Nations, Bonn, Germany

Makoto Inoue

Professor of Environmental Sociology, Faculty of Human Sciences, Waseda University and former Professor of Global Forest Environmental Studies, Graduate School of Agricultural and Life Sciences, the University of Tokyo, Japan

Foreword by Sir Robert T. Watson

Edward Elgar PUBLISHING

Cheltenham, UK • Northampton, MA, USA

Published by
Edward Elgar Publishing Limited
The Lypiatts
15 Lansdown Road
Cheltenham
Glos GL50 2JA
UK

Edward Elgar Publishing, Inc.
William Pratt House
9 Dewey Court
Northampton
Massachusetts 01060
USA

A catalogue record for this book
is available from the British Library

Library of Congress Control Number: 2016962578

This book is available electronically in the **Elgar**online
Economics subject collection
DOI 10.4337/9781785368455

MIX
Paper from responsible sources
FSC® C013604

ISBN 978 1 78536 844 8 (cased)
ISBN 978 1 78536 845 5 (eBook)

Typeset by Servis Filmsetting Ltd, Stockport, Cheshire
Printed and bound by CPI Group (UK) Ltd, Croydon, CR0 4YY

Contents

PART II CLIMATE RESILIENCE: SECTORAL PERSPECTIVES

PART III INCENTIVES, GOVERNANCE AND POLICY

Figures

Tables

Boxes

Contributors

Irina Arakelyan is Researcher, Oxford University Centre for the Environment, University of Oxford, UK.

Lint Barrage is Assistant Professor of Economics and Environmental Studies, Brown University, Providence, Rhode Island, USA.

Ian J. Bateman is Professor and Director of the Land, Environment, Economics and Policy (LEEP) Institute, University of Exeter, UK.

Patrick ten Brink is Senior Fellow, Institute for European Environmental Policy, Brussels, Belgium.

Carlo Carraro is Professor of Environmental Economics at Ca' Foscari University of Venice, Italy and Scientific Director, Fondazione Eni Enrico Mattei, Milan, Italy.

William W.L. Cheung is Associate Professor, Institute for the Oceans and Fisheries, University of British Columbia, Vancouver, Canada.

Robert Costanza is Professor of Public Policy, Crawford School of Public Policy, Australian National University, Canberra, Australia.

Philippe M. Cury is Europe-IRD Representative, Institut de Recherche pour le Développement (IRD), Brussels, Belgium.

Marinella Davide is Junior Researcher, Fondazione Eni Enrico Mattei, Venice, Italy.

Sabrina Dekker is Policy Analyst, Irish Solar Energy Association, Dublin, Ireland.

Anandi van Diepen is PhD student, Nohad A. Toulan School of Urban Studies and Planning,Portland State University, USA.

Yazan Elhadi is Monitoring, Evaluation and Research Officer of the Adaptation Consortium, Kenya National Drought Management Authority, Nairobi, Kenya.

Carlo Fezzi is Associate Professor, Department of Economics and

Management, University of Trento, Italy and Senior Lecturer, LEEP, University of Exeter, UK.

Iftekharul Haque is PhD candidate, School of International Development and Global Studies, University of Ottawa, Canada.

Amii R. Harwood is Lecturer, School of Environmental Sciences, University of East Anglia, Norwich, UK.

Ced Hesse is Principal Researcher, Climate Change Group, International Institute for Environment and Development, Edinburgh, UK.

Makoto Inoue is Professor of Environmental Sociology, Faculty of Human Sciences, Waseda University, Japan and former Professor of Global Forest Environmental Studies, Graduate School of Agricultural and Life Sciences, the University of Tokyo, Japan.

Craig Johnson is Professor, Department of Political Science, University of Guelph, Ontario, Canada.

Andrew A. Lovett is Professor, School of Environmental Sciences, University of East Anglia, Norwich, UK.

Kristy May is an Independent Researcher based in Toronto, Canada.

Kathleen A. Miller is Scientist, Climate Science and Applications Program, National Center for Atmospheric Research, Boulder, Colorado, USA.

Abrar Juhar Mohammed is JSPS Post-Doctoral Fellow, Laboratory of Global Forest Environmental Studies, the University of Tokyo, Japan.

Dominic Moran is Professor of Environmental Economics, Scotland's Rural College, Edinburgh, UK.

Konar Mutafoglu is Senior Policy Analyst, Institute for European Environmental Policy, Brussels, Belgium.

K.N. Ninan is Chairperson, Centre for Economics, Environment and Society, Bangalore, India and Co-Chair, Panel on Methodological Assessment of Scenarios and Models of Biodiversity and Ecosystem Services, Intergovernmental Platform on Biodiversity and Ecosystem Services, United Nations, Bonn, Germany.

Victor Orindi is National Coordinator of the Adaptation Consortium, Kenya National Drought Management Authority, Nairobi, Kenya.

Architesh Panda is Post-Doctoral Fellow, Social Sciences, International Rice Research Institute, Country Office, Vientiane, Lao PDR.

Anthony Patt is Professor of Human-Environment Systems, Institute for Environmental Decisions, ETH, Zurich, Switzerland.

Ramón Pichs-Madruga is Researcher and Director, Centre for World Economy Studies, Havana, Cuba.

Meenakshi Rao is Post-Doctoral Researcher, School of the Environment, Portland State University, USA.

Jean-Pierre Schweitzer is Policy Analyst, Institute for European Environmental Policy, Brussels, Belgium.

Vivek Shandas is Associate Professor, Nohad A. Toulan School of Urban Studies and Planning, Portland State University, USA.

Upasna Sharma is Assistant Professor, Humanities and Social Sciences, Indian Institute of Technology, Delhi, India.

Yvonne Su is PhD candidate in Political Science and International Development Studies, University of Guelph, Ontario, Canada.

U. Rashid Sumaila is Professor, Institute for the Oceans and Fisheries, and Liu Institute for Global Studies, University of British Columbia, Vancouver, Canada.

Travis Tai is PhD student, Institute for the Oceans and Fisheries, University of British Columbia, Vancouver, Canada.

David Timmons is Assistant Professor of Economics, University of Massachusetts, Boston, Massachusetts, USA.

Clem Tisdell is Professor Emeritus, School of Economics, The University of Queensland, Brisbane, Australia.

Jackson Voelkel is Researcher, Portland State University, USA.

Robert T. Watson is Chair, Intergovernmental Platform on Biodiversity and Ecosystem Services (IPBES), United Nations, Bonn, Germany and former Chair, Intergovernmental Panel on Climate Change (IPCC), United Nations.

Jamie Woollard is Associate, Ernst & Young, London, UK.

Anita Wreford is Researcher, Scotland's Rural College, Edinburgh, UK.

Foreword

Sir Robert T. Watson

One of the greatest challenges facing society is human-induced climate change. This challenge, which has been recognized by all governments in the world, led to the historic climate agreement in Paris in 2015. The Paris agreement called for holding the increase in global average temperature to well below 2 °C above pre-industrial levels and pursue efforts to limit the temperature increase to 1.5 °C above pre-industrial levels. The Paris agreement also called for an increase in the ability to adapt to the adverse impacts of climate change and foster climate resilience. It recognized the importance of developed countries providing financial assistance to developing countries to transition to a low-carbon economy and adapt to climate change.

The adverse effects of human-induced climate change on socio-economic sectors such as food and water security, human settlements, human health, and biodiversity and ecosystems have already been observed, and even if the 2 °C target is met, which is unlikely, then further adverse consequences will occur. Unless human-induced climate change is addressed urgently, many of the 17 Sustainable Development Goals will be undermined, especially in developing countries, which are most at risk owing to their limited technological and financial capability to adapt.

Understanding the synergies and trade-offs, the economic costs and benefits, and the social acceptability of climate adaptation and mitigation strategies is essential. What may be a beneficial policy for addressing climate change may be beneficial or detrimental to another issue such as biodiversity and ecosystems, or may lead to distributional effects, with one part of society benefiting while another part of society is adversely affected.

This book, *Building a Climate Resilient Economy and Society: Challenges and Opportunities*, is timely and identifies a range of options to adapt, reduce vulnerability and increase resilience to human-induced climate change for both terrestrial and marine systems. It addresses key sectors such as agriculture, fisheries, water quantity and quality, and coastal cities, and key issues such as terrestrial and marine biodiversity and small island states. It also addresses key issues associated with mitigation, including carbon pricing, economic implications of climate policies, financing at local levels, and REDD+. I would like to congratulate the editors and authors for bringing out this book, which I am sure will receive wide attention.

Preface

Climate change poses an immense challenge to governments, societies and entities. Evidence collated by the Intergovernmental Panel on Climate Change (IPCC) depicts a grim picture of the trends in climate variables. These indicate an unprecedented rise in global average land and ocean temperatures, the melting of glaciers, rising sea levels, increased frequency and intensity of natural calamities, and so on. If current lifestyles and economic activities continue, average global air temperature is likely to rise by between 2 and 4.5 °C or even more by the end of this century. If this happens it will have disastrous consequences on ecosystems, species, lives and livelihoods. Poor and marginalized people, developing countries and small island states are most vulnerable to the risks posed by climate change. Climate change may also lead to transboundary conflicts (e.g. over access to and use of shrinking water resources) or movement of environmental refugees. Greenhouse gas (GHG) emissions, especially those of CO_2 caused by anthropogenic factors, are held responsible for global warming. Atmospheric concentration of carbon dioxide crossed 400 ppm in 2015 compared with 280 ppm during the pre-industrial period. If these trends persist there is a 50 per cent risk of a global temperature increase of 5 °C by the end of this century. Temperature rise also won't be even over the globe, and hot and semi-arid regions might even witness rises of above 5 °C. The Stern Review of 2007 suggests that, with a 5–6 °C warming, an average of 5 to 10 per cent loss in global GDP is possible, with poor countries suffering losses of more than 10 per cent of GDP. Keeping the average rise in global temperature to below 2 °C above pre-industrial levels, and if possible below 1.5 °C above pre-industrial levels, has been accepted as a goal by 195 countries that approved the Paris Climate Agreement in December 2015. However, prospects for this are dim given the current pledges for reducing emission levels.

In this context it is important for governments, societies and entities to build resilience to face the risks posed by climate change and extreme weather events. This book seeks to address the challenges and opportunities in building a climate resilient economy and society. The chapters included in this volume cover a broad range of topics, such as: vulnerability, adaptation and resilience; challenges and prospects for building climate resilience in different sectors such as agriculture, marine ecosystems, urban areas, drought prone areas, and energy; carbon pricing and financing; REDD+; climate policies and governance; and so on. The chapters include those that have a global or regional focus as well as case studies drawn from a cross-section of countries in Africa, Asia, Europe and North America. The contributors to this book are leading experts from around the world who have made a significant contribution to the literature in this area. Some of them have also been part of IPCC Assessments. For convenience, the discussion in the book is organized in three parts: I, 'Vulnerability, adaptation and resilience'; II, 'Climate resilience: Sectoral perspectives'; and III, 'Incentives, governance and policy'.

At the outset we would like to express our sincere thanks to all the contributing authors for readily responding to our invitation to contribute a chapter to this book, and for putting up with our frequent emails reminding them about the timely submission of their chapters and revised chapters in the light of reviewers' comments, as well as responding to our queries and clarifications. But for their unstinted cooperation and support this volume would not have been possible.

The book has undergone a rigorous review process. Besides the editors, most chapters were reviewed by two or more reviewers. We are extremely grateful to the following contributing authors for sparing their valuable time and giving several useful comments on the chapters. They are: Patrick ten Brink, Anandi van Diepen, Carlo Fezzi, Ced Hesse, Craig Johnson, Abrar Juhar Mohammed, Dominic Moran, Architesh Panda, Ramón Pichs-Madruga, Vivek Shandas, U. Rashid Sumaila, David Timmons, Clem Tisdell and Jackson Voelkel. In addition to the contributing authors, we are also grateful to the following external reviewers for reviewing the chapters and giving several suggestions and comments which were useful to the authors for improving the rigor and quality of their chapters. They are: Luke Brander (IVM, Amsterdam), M.G. Chandrakanth (University of Agricultural Sciences, Bangalore), Johane Dikgang (University of Johannesburg), Charles Howe (Colorado State University), Rohit Jindal (MacEwan University, Canada), Elumalai Kannan (JNU, New Delhi), K.S. Kavi Kumar (MSE, Chennai), Andreas Kontoleon (Cambridge University), Parmod Kumar (ISEC, Bangalore), Eric Mungatana (University of Pretoria), Sunil Nautiyal (ISEC, Bangalore), Jyothi Parikh (Irade, New Delhi), M.V. Ramana (Princeton University), Jyothis Satyapalan (CESS, Hyderabad) and David Simpson (US EPA, Washington).

We are also grateful to the following authors and publishers (Nature Publishing Group and Elsevier) for permission to publish the following articles in this book:

Fezzi, C., A.R. Harwood, A.A. Lovett and I.J. Bateman (2015), 'The environmental impact of climate change adaptation on land use and water quality', *Nature Climate Change*, **5**, 255–260, doi:10.1038/nclimate2525.
Panda, A., U. Sharma, K.N. Ninan and A. Patt (2013), 'Adaptive capacity contributing to improved agricultural productivity at the household level: Empirical findings highlighting the importance of crop insurance', *Global Environmental Change*, **23**, 782–790.
Costanza, R. (2015), 'Claim the sky!', *Solutions*, January–February, 18–21.

We are most grateful to Sir Robert T. Watson for finding time to write a foreword to this book. His dedication and leadership in the IPCC earlier and now in the IPBES have been an inspiration for many.

Finally we would like to express our sincere thanks to Edward Elgar Publishing for readily agreeing to publish this book and to the editors and entire Edward Elgar staff for the unstinted support provided for publishing this book.

K.N. Ninan and Makoto Inoue

Abbreviations and acronyms

ACS	American Community Survey
ADB	Asian Development Bank
ADS-E	Anglican Development Services–Eastern
AENOR	Asociación Española de Normalización y Certificació
AF	acre-foot; or agroforestry
ALDEF	Arid Lands Development Focus
AMAP	Arctic Monitoring and Assessment Programme
AOSIS	Alliance of Small Island Developing States
AR4	Fourth Assessment Report (of IPCC)
AR5	Fifth Assessment Report (of IPCC)
ASAL	arid and semi-arid lands
BAU	business as usual
BCCRF	Bangladesh Climate Change Resilience Fund
BCCSAP	Bangladesh Climate Change Strategy and Action Plan
BERSMP	Bale Eco-Region Sustainable Management Programme
BMI	body mass index
BPL	below poverty line
BPS	Bureau of Planning and Sustainability
BTA	border tax adjustment
CA	conservation agriculture
CAES	compressed air energy storage
CAP	climate action plan
CAPC	county adaptation planning committee
CBG	census block group
CCBA	Climate Community and Biodiversity Alliance
CCCF	County Climate Change Fund
CCE	cost of conserved energy
CCF-E	Climate Change Forum–Ethiopia
CDMs	county directors of meteorology
CDWR	California Department of Water Resources
CE	conserved energy
CEX	Consumer Expenditure Survey

CF	capacity factor
CGE	computable general equilibrium
CGIAR	Consortium of International Agricultural Research Centers
CHP	combined heat and power
CIDP	County Integrated Development Plan
CIS	climate information services
CLF	Coalition for a Livable Future
CO_2	carbon dioxide
COP	Conference of the Parties
CRGE	Climate-Resilient Green Economy
CSA	climate smart agriculture
CSP	concentrating solar plants
CVP	Central Valley Project
CVPIA	Central Valley Project Improvement Act
D	demand curve
DAP	Detailed Area Plan
DC	direct current
DFID	Department for International Development
DMDP	Dhaka Metropolitan Development Plan
DWR	Department of Water Resources
EACC	economics of adaptation to climate change
EAF	ecosystem approach to fisheries
EC	European Commission
EEA	European Environment Agency
EEZ	exclusive economic zone
EIA	Energy Information Administration
EPA	Environmental Protection Authority
ESA	Endangered Species Act
ESRC	Economic and Social Research Council
ETR	environmental tax reforms
ETS	emissions trading scheme
EU	European Union
EU ETS	EU Emissions Trading System
EUR	euro
EV	equivalent variation
FAO	Food and Agriculture Organization
FAR	First Assessment Report (of IPCC)
FCPF	Forest Carbon Partnership Facility
FGM	farm gross margin

FIP	Forest Investment Programme
FPIC	free prior and informed consent
GB	Great Britain
GCF	Green Climate Fund
GDP	gross domestic product
GEF	Global Environment Facility
GFDRR	Global Facility for Disaster Reduction and Recovery
GHG	greenhouse gas
GIS	geographical information system
GM&T	Gazprom Marketing and Trading
GQA	General Quality Assessment
Gt	gigatonne
HCMC	Ho Chi Minh City
HDI	Human Development Index
HFCs	hydrofluorocarbons
IAMs	integrated assessment models
IBRD	International Bank for Reconstruction and Development
ICCCF	Isiolo County Climate Change Fund
IDESAM	Institute for Conservation and Sustainable Development of Amazonas
IEA	International Energy Agency
IFPRI	International Food Policy Research Institute
IIED	International Institute for Environment and Development
INDC	intended nationally determined contribution
IOA	input–output analysis
IPBES	Intergovernmental Science-Policy Platform on Biodiversity and Ecosystem Services
IPCC	Intergovernmental Panel on Climate Change
K	capital cost
KACP	Kenya Agriculture Carbon Project
KMD	Kenya Meteorological Department
kWh	kilowatt hours
LCOE	levelized cost of energy
LDC	least developed countries
LEDs	light emitting diodes

M&E	monitoring and evaluation
MA	Millennium Ecosystem Assessment
MAC	marginal abatement cost
MAES	mapping of ecosystem services
MAF	million acre-feet
MC	marginal cost
MCF	marginal cost of public funds
Mm^3	million cubic metres
MMU	Manchester Metropolitan University
$MMtCO_2e$	million metric tons of carbon dioxide equivalent
MoA	Ministry of Agriculture
MoEF	Ministry of Environment and Forests
MOP	meeting of the parties
MPAs	marine protected areas
MRAs	marine recreational activities
MRV	measuring, reporting and verification
$MtCO_2e$	metric ton of carbon dioxide equivalent
MW	megawatt
MWD	Metropolitan Water District
MWh	megawatt hour
NAPA	national adaptation programme of action
NCRC	Nature Conservation Research Centre
NDMA	National Drought Management Authority
NGO	non-governmental organization
NIE	national implementing entity
NO_3	nitrate
NOAA	National Oceanic and Atmospheric Administration
NOK	Norwegian krone
NRS	National REDD+ Secretariat
NSDP	net state domestic product
NSS	National Sample Survey
ODA	official development assistance
OECD	Organisation for Economic Co-operation and Development
OFWE	Oromia Forest and Wildlife Enterprise
P	maximum power output; or phosphorus
PAMs	policies and measures
PFM	participatory forest management
PIT	personal income tax

ppm	parts per million
PV	photovoltaic
PVID	Palo Verde Irrigation District
r	interest rate
R&D	research and development
RAJUK	*Rajdhani Unnayan Kartripakkha* (Capital Development Authority of the Government of Bangladesh)
RAP	Resource Advocacy Programme
RED	reducing emissions from deforestation
REDD+	reducing emissions from deforestation and forest degradation plus
REHAB	Real Estate Housing Association of Bangladesh
RMG	ready-made garment
R-PIN	Readiness Plan Idea Note
R-PP	readiness preparation plan
RRI	Rights and Resources Initiative
SAWPA	Santa Ana Water Project Authority
SBI	Subsidiary Body for Implementation
SBSTA	Subsidiary Body for Scientific and Technological Advice
SD	sustainable development
SDGs	Sustainable Development Goals
SI	sustainable intensification
SIDS	small island developing states
SM	supplementary material
SMC	social marginal cost
SRES	Special Report on Emission Scenarios (of the IPCC)
SWP	State Water Project
SWRCB	State Water Resources Control Board
T	life of project
TAMD	Tracking Adaptation and Measuring Development
TAR	Third Assessment Report (of IPCC)
TEEB	The Economics of Ecosystems and Biodiversity
ToC	theory of change
UHI	urban heat island
UK	United Kingdom
UKCIP	United Kingdom Climate Impacts Programme
UN	United Nations
UNCBD	United Nations Convention on Biological Diversity

UNCCD	United Nations Convention to Combat Desertification
UNDESA	United Nations Department of Economic and Social Affairs
UNDP	United Nations Development Programme
UNEP	United Nations Environment Programme
UNESCO	United Nations Educational, Scientific and Cultural Organization
UNFCCC	United Nations Framework Convention on Climate Change
UNISDR	United Nations Office for Disaster Risk Reduction
UPI	Urban Planning Institute
US	United States
USA	United States of America
USD	US dollars
USDA	United States Department of Agriculture
USGS	United States Geological Survey
VCAPS	Vietnam Climate Adaptation Partnership
VCS	Verified Carbon Standard
VMT	vehicle miles travelled
VSD	vulnerability scoping diagram
WAPC	ward adaptation planning committee
W_{DC}	watts direct current
WFD	Water Framework Directive
WG	Working Group
WHO	World Health Organization

1. Introduction

K.N. Ninan and Makoto Inoue

BUILDING A CLIMATE RESILIENT ECONOMY AND SOCIETY

Climate change poses a great challenge to governments, societies and entities. Finding ways to transit to a climate resilient economy and society therefore assumes great importance. Average global atmospheric CO_2 concentration has increased from the pre-industrial level of 280 parts per million (ppm) to 400.8 ppm in 2015. The annual growth rate of atmospheric carbon dioxide has roughly tripled from 0.6 ppm per year in the early 1960s to an average of 2.1 ppm during the past ten years, and further rose to a high of 3.05 ppm in 2015 according to observations recorded at the Mauna Loa Observatory, Hawaii (NOAA, 2016). Global CO_2 emissions rose from about 22.5 billion tons in 1990 to 35.7 billion tons in 2014 (Table 1.1). While developed countries such as the USA, EU 28, Japan, Australia and Canada put together accounted for about 30.7 per cent of the total global CO_2 emissions in 2014, China's share alone was about 29.7 per cent (Table 1.1). Although fast growing economies such as India have witnessed a rapid rise in CO_2 emissions recently, in per capita terms India's CO_2 emissions (1.8 tons per capita) are still far below those prevailing in developed countries, which ranged between 10 and 17.3 tons per capita in 2014 (Table 1.1).

The Fifth Assessment Report of the Intergovernmental Panel on Climate Change (IPCC) presents a grim picture of the trends in climatic variables. Its key findings, amongst others, note as follows (IPCC, 2014a):

- Human influence on the climate system is clear, and recent anthropogenic emissions of greenhouse gases (GHGs) are the highest in history. Anthropogenic GHG emissions have increased since the pre-industrial era, driven largely by economic and population growth, and are now higher than ever.
- Each of the last three decades has been successively warmer at the Earth's surface than any preceding decade since 1850. The period

Table 1.1 Share of selected countries in CO_2 emissions and per capita CO_2 emissions (1990 and 2014)

Country	Share in CO_2 emissions in % 1990	Share in CO_2 emissions in % 2014	CO_2 emissions in tons per capita 1990	CO_2 emissions in tons per capita 2014
China	10.7	29.7	2.1	7.6
USA	22.2	14.8	19.6	16.5
EU 28	19.1	9.5	9.2	6.7
India	3.1	6.4	0.8	1.8
Russia	10.7	5.0	16.1	12.4
Japan	5.3	3.6	9.6	10.1
Canada	1.8	1.7	16.2	15.9
Brazil	0.9	1.4	1.5	2.5
Australia	1.3	1.1	16.1	17.3
South Africa	1.3	1.1	7.3	7.4
Global total	22.5 billion tons of CO_2	35.7 billion tons of CO_2		

Source: Olivier et al. (2015).

from 1983 to 2012 was likely the warmest 30-year period of the last 1400 years in the Northern Hemisphere where such assessment is possible (*medium confidence*).

- The globally averaged combined land and ocean surface temperature data as calculated by a linear trend show a warming of 0.85 (0.65 to 1.06) °C over the period 1880 to 2012.
- Global mean sea level rose by 0.19 m (0.17 to 0.21 m) between 1901 and 2010. The rate has been higher since the mid-19th century as compared to the previous two millenniums (*high confidence*).
- Over the period 1992 to 2011, the Greenland and Antarctic ice sheets lost mass (*high confidence*), likely at a greater rate over 2002 to 2011.
- Surface temperature is projected to rise over the 21st century under all assessed emission scenarios. It is very likely that heat waves will occur more often and last longer, and that extreme precipitation events will become more intense and frequent in many regions. The ocean will continue to warm and acidify, and global mean sea level will continue to rise.
- A large fraction of species face increased risk owing to climate change during and beyond the 21st century, especially as climate change interacts with other stressors (*high confidence*). The Fourth

> Assessment Report of the IPCC notes that globally approximately 20 to 30 per cent of species assessed so far are *likely* to be at increased risk of extinction if increases in global average warming exceed 1.5 to 2.5 °C (relative to 1980–1999 temperature levels). As global average temperature increase exceeds about 3.5 °C, model projections suggest significant extinctions, that is, 40 to 70 per cent of species assessed (IPCC, 2007).

If these trends continue, which is most likely, and exacerbate further in the absence of efforts to stabilize emissions, it will have disastrous consequences on human and natural systems. Economic estimates of the costs of climate change vary widely across different studies, countries and regions depending upon the assumptions made regarding future emission scenarios, period of analysis, discount rate used, and so on. The Stern Review (Stern, 2007) suggests that, if we don't act, the overall costs and risks of climate change will be equivalent to losing at least 5 per cent of global gross domestic product (GDP) each year, now and for ever. If a wider range of risks and impacts is considered, the estimates of damage could rise to 20 per cent of GDP or more (Stern, 2007). However, the Stern Review estimates were widely criticized. Tol and Yohe (2006), for instance, note that the estimates in the Stern Review were based on existing published literature and were on the lower side (losses below 5 per cent of GDP). Further the high valuation of climate change impacts in the Stern Review were due to the use of a very low discount rate, risk that is double counted, and vulnerability that is assumed to be constant over very long periods of time (two or more centuries, to be exact). A review of estimates by different researchers notes that, based on assumptions made regarding future emission scenarios and for other parameters, for warming levels of between 1 and 3 °C the impact on GDP ranged from 2.5 per cent to −4.8 per cent (Tol, 2009). These estimated losses in GDP were much higher for Africa, Asia and South America, ranging from −2.6 per cent to −23.5 per cent across different studies (Tol, 2009). A recent study by Ahmed and Suphachalasai (2014) notes that, without global deviation from a fossil fuel intensive path, South Asia could lose an equivalent of 1.8 per cent of its annual GDP by 2050, which will progressively increase to 8.8 per cent by 2100 on the average under a business as usual (BAU) scenario. The Fifth Assessment Report of the IPCC notes that economic impact estimates completed over the past 20 years vary in their coverage of subsets of economic sectors and depend on a large number of assumptions, many of which are disputable, and many estimates do not account for catastrophic changes, tipping points, and many other factors. With these recognized limitations, the incomplete estimates of global annual economic losses for additional temperature increases of ~2 °C are between 0.2 and 2.0 per cent of income (±1 standard deviation around the mean) (*medium evidence*, *medium agreement*). Losses are *more likely than not* to be greater, rather than smaller, than this range (*limited evidence*, *high agreement*). Additionally, there are large differences between and within countries. Losses accelerate with greater warming (*limited evidence*, *high agreement*), but few quantitative estimates have been completed for additional warming around 3 °C or above (IPCC, 2014b).

To address the risks (e.g. hazards and extreme weather events) posed by climate change we need to build the resilience of people, areas and sectors to cope with these risks. Climate resilience can be generally defined as the capacity of a socio-ecological system to: (1) absorb stresses and maintain function in the face of external stresses imposed upon it by climate change; and (2) adapt, reorganize and evolve into more desirable configurations that improve the sustainability of the system, leaving it better prepared for future climate change impacts (Folke, 2006; Nelson et al., 2007). The IPCC (2014b) defines 'resilience' as 'the capacity of social, economic and environmental systems to cope with a hazardous event or trend or disturbance, responding or reorganizing in ways that maintain their essential function, identity and structure, while also maintaining the capacity for adaptation, learning and transformation'. To implement policies to enhance climate resilience we need to assess the vulnerability of people, areas and sectors to the risks posed by climate change. Vulnerability encompasses three components, namely adaptive capacity, exposure and sensitivity. Different approaches such as historical narratives, geographical information system (GIS) and mapping techniques, agent-based modelling and an indicator-based approach have been used to assess vulnerability to climate change (Panda, 2016). Of these, indexes of vulnerability based on socio-economic, biophysical and other parameters have been popular and used by researchers, for example, to assess the vulnerability of coastal communities in Vietnam (Adgers, 1999) and the vulnerability to climate change of people living in drought-prone areas in Odisha, India (Panda, 2016). Poor and marginalized people, indigenous communities, and coastal communities, ecologically fragile areas such as hill and dry regions, coastal areas, and agriculture and the primary sector are most vulnerable to the adverse effects of climate change. Adaptation and mitigation are complementary strategies for reducing and managing the risks of climate change (IPCC, 2014a). Mitigation implies that the causes of climate change are addressed by reducing GHG emissions, whereas adaptation requires that the effects of climate change are dealt with by coping with their negative impacts. Substantial emission reductions over the next few decades can reduce climate risks in the 21st century and beyond, increase prospects for effective adaptation, reduce the costs and challenges of mitigation in the longer term and contribute to climate resilient pathways for sustainable development (IPCC, 2014a). Adaptation and mitigation responses are underpinned by common enabling factors. These include effective institutions and governance, innovation and investments in environmentally sound technologies and infrastructure, sustainable livelihoods and behavioural and lifestyle choices (IPCC, 2014a).

The historic climate pact approved by 195 countries attending the 21st Conference of the Parties (COP) to the United Nations Framework Convention on Climate Change (UNFCCC) held in Paris (30 November to 11 December 2015) provides a roadmap to the global community to inch towards a climate resilient economy and society. The agreement, which comes into effect from the year 2020, resolved to limit the global temperature increase to well below 2 °C above pre-industrial levels and pursue efforts to limit temperature rise to 1.5 °C above pre-industrial levels if possible. It further agreed that, while rich countries would engage in absolute emission reductions, developing countries would enhance mitigation efforts. The accord also agreed to periodic monitoring and verification of all countries' GHG inventory and emission reductions. It agreed to mobilize climate finance of US$ 100 billion per year by 2020 to 2025 and a new higher goal thereafter to assist developing countries in reducing emissions and adapting to climate change. Following its ratification by the requisite number of countries in accordance with Article 21, Para 1, the Paris Climate Agreement came into force with effect from 4 November 2016 (UNFCCC, 2016).

Need for Building Climate Resilience

An obvious question that arises is why we need to build climate resilience. The risks arising from climate change are many, such as unforeseen and extreme weather events like heat waves, typhoons and cyclones, coastal and river flooding, prolonged droughts, and so on. These can have adverse economic, social and environmental consequences and affect human well-being and the overall quality of life. For instance, it is estimated that between 1980 and 2014 weather-related losses were estimated at US$ 3.3 trillion (2014 values adjusted for inflation based on country CPI), of which only about 28.5 per cent was insured, that is, US$ 0.94 trillion (Munich RE, 2015). Some 88 per cent of these reported disasters (total 21,700 disaster events), 78 per cent of total losses from all disasters (US$ 4.2 trillion expressed in 2014 values) and 49 per cent of lives lost (total lives lost from all disasters being 1.74 million) were caused by weather extremes (Munich RE, 2015). It is also estimated that large coastal cities alone could face combined annual losses of US$ 1 trillion from flooding by mid-century (cited in Royal Society, 2014). Hence we need to be well prepared to cope with such eventualities. Further rising temperatures, shrinking of mountain glaciers and variations in other climatic parameters have led to diseases or their carriers reaching higher elevations or the emergence of new diseases such as dengue. Increasingly, these factors are driving new and unpredictable patterns in the spread of common water- and vector-borne diseases, with conventional protocols for monitoring and managing them under increasing strain (Dayal, 2014). A recent study by the Asian Development Bank (ADB) suggests that under low, medium and high emission scenarios (B1, A1B and A2 SRES scenarios of the IPCC) the number of people in India who would be affected by malaria by 2050 is estimated to range between 1.71 million and 3 million, and the number of deaths due to malaria is estimated to range between 0.01 million and 0.12 million (Ahmed and Suphachalasai, 2014). Similarly, the number of people in India who will be affected by diarrhoea by 2050 is estimated to range between 24.3 million and 42.4 million under low, medium and high emission scenarios (Ahmed and Suphachalasai, 2014). Building climate resilience is therefore essential to tackle these emerging public health challenges and rising health costs. Renewable surface water and groundwater resources in most dry subtropical regions are projected to reduce owing to climate change, intensifying competition for water among sectors (IPCC, 2014a). This will aggravate conflicts between households, sectors and regions over access to and use of shrinking water resources. Climate change may also lead to transboundary conflicts (e.g. over access to and use of water resources) or movement of environmental refugees (Cammack, 2007). Rising demand for food due to rising population and incomes coupled with declining production of staples such as wheat, rice and maize due to the adverse effects of climate change will undermine food security and affect the poor and vulnerable people. Mainstreaming 'resilience' into development plans is therefore critical to tackling poverty, inequality, ill health and poor sanitation, as well as realizing the Sustainable Development Goals (SDGs), since poor and marginalized people and poor countries that have low adaptive capacity are most vulnerable to the adverse consequences of climate change. Improving risk management can lead to larger gains in development and poverty reduction (World Bank, 2013a). Implementing climate and disaster resilient development plans may entail high start-up costs, but they are cost effective in the long run and could reduce the costs due to disaster. Early warning systems, better preparedness and improved safety codes have proven to be cost effective, save human lives and protect public and private investment (World Bank, 2013b). For instance, Cyclone Phailin, which struck Odisha and Andhra

Pradesh states on the east coast of India in October 2013, resulted in 40 deaths only, compared to the 10000 who perished during a similar event in 1999 in the absence of such early warning systems (World Bank, 2013b). Building climate resilience will also be helpful in coping with uncertainties arising from the complex and dynamic interactions between climate change and other parameters such as health.

Approaches to Build Climate Resilience

There are a number of approaches to build climate resilience. These are ecosystem-based, engineering-based, hybrid (i.e. a combination of ecosystem- and engineering-based), social and behavioural approaches (Royal Society, 2014). Table 1.2 gives examples of using these adaptation options to address extreme weather events such as heat waves, coastal flooding, droughts and river flooding. For instance, an ecosystem-based approach involves building resilience through green solutions such as afforestation or reforestation to capture carbon, encouraging coastal afforestation and maintaining or re-establishing mangroves to reduce the impact of coastal flooding and storm surges, maintaining and increasing tree cover in cities and towns to reduce the consequences of heat waves, promoting agro-forestry in drought-prone areas, and so on. An engineering-based approach includes investing in human-made structures such as building sea walls or coastal barrages to cope with sea level rise and coastal erosion, building cyclone shelters in cyclone-prone areas, and so on. A hybrid approach involves combining ecosystem- and engineering-based approaches to address weather extremes and disasters. For example, to build resilience against sea erosion, sea level rise and storm surges one could construct sea walls or coastal barrages along with planting trees in coastal boundaries prone to such problems. A social approach involves building and strengthening social networks and community-led responses and actions to adapt to climate change. A behavioural approach involves transforming the mindset, attitudes and values of people in order to change their lifestyles and take proactive actions that will help address the adverse effects of climate change. These are not mutually exclusive approaches and could complement each other. For instance, community-led responses and actions to address the ill effects of climate change would entail a change in the attitudes and values of individuals and communities. These approaches have their advantages and disadvantages. For instance, an engineering-based approach requires considerable funds and capital to be invested in building human-made structures such as sea walls, cyclone shelters, and so on to mitigate the adverse impacts of climate change, whereas an ecosystem-based approach may be a low cost option. But the

Table 1.2 Examples of different ecosystem- or engineering-based or hybrid (i.e. combination of ecosystem- and engineering-based) options to reduce the impact of four types of extreme weather events

Extreme weather events	Examples
Heat waves	Afforestation; maintenance of existing vegetation; green roofs, vertical greening systems; white roofs; urban planning, grid design, etc.; air conditioning; insulation.
Coastal flooding	Maintenance of natural reefs (coral/oyster); maintenance of mangroves; mangrove planting and re-establishment; maintenance and creation of salt marshes, wetlands, inter-tidal ecosystems; maintenance of coastal, forest and other ecosystems; coastal re-vegetation and afforestation (above inter-tidal zone); beach and dune nourishment; artificial reefs; dykes, levees; coastal barrages.
Drought	Removal of thirsty invasive plant species; reforestation; forest conservation; agroforestry; breeding drought resilient crops and livestock; sustainable agroecosystem management practices; soil and water conservation; reservoirs, ponds and other water storage; wells; irrigation; inter-basin water transfer; waste-water recycling.
River flooding	Re-establishment of floodplains, green rivers; catchment afforestation, increased vegetation cover; maintenance of existing catchment vegetation; planting of riparian buffers; changes to catchment agricultural land management; natural flood management; stream habitat restoration; dams; drains, dykes, levees, sluices, pumps; dredging; sustainable urban drainage systems; canalization of urban streams.

Source: Royal Society (2014: 60–61).

ecosystem-based approach may have high opportunity costs such as the need for land to raise afforestation/reforestation, and so on (Royal Society, 2014). An engineering-based approach too could have high opportunity costs, since funds that would otherwise be used for development activities must be diverted to invest in human-made structures to combat sea level rise, coastal erosion, and so on. An ecosystem-based approach can have several co-benefits (SCBD, 2009). Table 1.3 illustrates the adaptive

Table 1.3 Examples of the co-benefits of ecosystem-based adaptation measures in response to climate change

Adaptation measure	Adaptive function	Co-benefits			
		Social and cultural	Economic	Biodiversity	Mitigation
Mangrove conservation	Protection against storm surges, sea level rise and coastal inundation.	Provision of employment opportunities (e.g. fisheries and prawn cultivation); contribution to food security.	Generation of employment opportunities to local communities through marketing of mangrove products (e.g. fish, dyes, medicines).	Conservation of species that live or depend on mangroves.	Conservation of carbon stocks, both above and below ground.
Forest conservation and sustainable forest management	Maintenance of nutrient and water flow; prevention of landslides.	Opportunities for recreation, culture, and protection of indigenous people and local communities.	Potential generation of income through ecotourism, recreation, sustainable logging.	Conservation of forest plant and animal species.	Conservation of carbon stocks; reduction of emissions from deforestation and degradation.
Restoration of degraded wetlands	Maintenance of nutrient and water flow, quality, storage and capacity; protection against flood or storm inundation.	Sustained provision of: livelihood, recreation, employment opportunities.	Increased livelihood generation, potential revenue from recreational activities; sustainable logging of planted trees.	Conservation of wetland flora and fauna, and stop-over sites for migratory species.	Reduced emissions from soil carbon mineralization.
Establishment of diverse agroforestry systems in agricultural lands	Diversification of agricultural production to cope with changed climate conditions.	Contribution to food and fuelwood security.	Generation of income from sale of timber, firewood and other products.	Conservation of biodiversity in agricultural landscapes.	Carbon storage in above and below ground biomass and soils.

Table 1.3 (continued)

Adaptation measure	Adaptive function	Co-benefits			
		Social and cultural	Economic	Biodiversity	Mitigation
Conservation of agrobiodiversity	Provision of gene pool for crop and livestock adaptation to climate change.	Enhanced food products; diversification of food products; conservation of local and traditional knowledge and practices.	Possibility of agricultural incomes in difficult environments; environmental services such as bees for pollination of cultivated crops.	Conservation of genetic diversity of crop varieties and livestock breeds.	
Conservation of medicinal plants	Local medicines available for health problems resulting from climate change or habit degradation (e.g. malaria, diarrhoea, etc.).	Local communities have an independent and sustainable source of medicines; maintenance of local knowledge and traditions.	Potential source of income for local people.	Enhanced medicinal plant conservation; local traditional knowledge recognized and protected.	Environmental services such as bees for pollination of cultivated crops.
Sustainable management of grasslands	Protection against floods; nutrient storage; maintenance of soil structure.	Recreation and tourism.	Generate income for local communities through products from grass (e.g. brooms).	Forage for grazing animals; provide diverse habitats for animals that are predators and prey.	Maintenance of soil carbon; storage of soil carbon.

Source: SCBD (2009: 43–44, table 2.3). Reproduced with permission from the SCBD, Montreal.

function and co-benefits of various adaptation options to address climate change. For instance, conservation of mangroves, which helps to protect against storm surges, sea level rise, coastal inundation, and so on, has several co-benefits, such as generating employment opportunities in fisheries, prawn cultivation, conserving mangrove-dependent species, carbon sequestration, providing a nursery for fish, and so on. Using a portfolio of approaches to address the multiple risks posed by climate change may yield greater benefits and win–win outcomes.

Challenges and Opportunities for Building Climate Resilience

Ensuring the implementation of the Paris Climate Agreement according to the agreed pledges, targets and timelines poses the biggest challenge in building climate resilience. This calls for cooperation and action at different scales (local, national and international) and among different stakeholders such as the state, local communities, civil society, and different interest groups including the private sector. Moreover the United Nations Environment Programme (UNEP) notes that even if all the emission reduction pledges made by countries in their intended nationally determined contribution (INDC) submitted to the UNFCCC are taken into account it still falls short of the emission reductions required to ensure a temperature rise of below 2 °C or a desirable 1.5 °C over pre-industrial levels by the year 2100 as resolved in the Paris Climate Agreement (UNEP, 2015). Giving favoured access to energy-efficient and other clean technologies and funding to transit to a low carbon economy and society is critical for developing countries which are still heavily dependent on fossil fuel-led growth. According to the Paris Climate Agreement developed countries pledged to raise funds to the tune of US$ 100 billion per annum by 2020 to help developing countries to adopt clean technologies and transit to a low carbon economy. But this appears to be far below the funding required for the mitigation and adaptation needs of developing countries. For instance, India's INDC submitted to the UNFCCC notes that between 2015 and 2030 India alone needs US$ 206 billion (at 2014–2015 prices) for implementing adaptation actions in agriculture, forestry, fisheries, infrastructure, water resources, and so on (UNFCCC, 2015). Estimates by India's NITI Aayog (the successor to India's former Planning Commission) indicates that the mitigation costs for moderate low carbon development would be around US$ 834 billion till 2030 at 2011 prices (UNFCCC, 2015). To give a sense of the funds needed vis-à-vis the pledges made in the Paris Climate Agreement it is worth noting that in the year 2015 US annual military expenditure was around US$ 598.5 billion; similarly, the annual global market for luxury goods was estimated at US$ 1 trillion in 2015, and for luxury cars about US$ 458 billion (Ninan, 2016). Addressing development needs as well as ensuring climate action is especially challenging for developing countries and new emerging economies which are struggling to reduce poverty levels and meet basic needs, but are still dependent on fossil fuel-led growth. Fossil fuel-dependent industries and sectors have concerns about losing their competitiveness and markets vis-à-vis the renewable energy sector. International trade and trading policies will increasingly be driven by these green constraints. Building a political consensus and will to shift from a carbon-subsidizing regime to a carbon-taxing regime is another challenge facing many countries. There is also a need for building the capacity of different actors and institutions, including different layers of government to cope with the challenges posed by climate change. Financial reforms including in the insurance sector are needed to cover the risks posed by climate change, especially owing to rising costs and strains on governmental and household finances arising from weather-related extremes and disasters, as noted earlier. Rising incomes and global population, which is projected to rise to 9.3 billion by the year 2050 from 7 billion in 2011 according to the 2010 medium projections of the UN, pose further challenges in promoting climate resilient policies (Royal Society, 2014).

The recent international agreements on climate change, disaster reduction and phasing out of hydrofluorocarbons (HFCs) and their implementation present an opportunity to develop a coherent strategy to build resilience to extreme weather locally, nationally and internationally, and to shape plans, responses and outcomes for decades to come. There is also growing recognition about the role of the natural environment in reducing hazards, and recognition of the importance of mobilizing community support (Royal Society, 2014). For instance, evidence suggests that, in the aftermath of the 2004 tsunami, areas with intact mangrove forests could withstand the huge devastation caused by the tsunami in South and South-East Asia, whereas in areas where mangroves and other coastal habitats had been destroyed, often illegally, the waves could penetrate far inland, destroying homes, inundating farmland and washing away people and livelihoods (EJF, 2006). SCBD (2009) notes that integrating biodiversity and provision of ecosystem services is cost effective, and can generate social, economic and cultural co-benefits and help maintain resilient systems. Scientific and technological advances, improvements in forecasting and warning systems, and so on have enabled states and communities to be better prepared than ever before to cope with weather extremes and disaster events. Satellite imagery and high resolution data and use of more sophisticated models provide policymakers with unprecedented access to reliable data with which to make better informed decisions (Royal Society, 2014). Smart phones are increasingly being used to transmit the latest weather forecasts to farmers, farming villages and village level institutions, which enables them to make timely decisions on the commencement date for sowing crops or to prepare for weather extremes such as droughts or floods. New crop varieties and technologies provide an opportunity to make agriculture more sustainable and climate resilient as well as meet the needs of a growing population with better incomes. Improved awareness of the risks of global warming along with scientific evidence has created a better environment for climate action and resilient-building measures at the local, national and international scales as well as among different stakeholders. Integration of indigenous, local and scientific knowledge will be helpful in taking advantage of different knowledge systems to enhance climate and disaster resilience. There is also growing availability of funding and initiatives for climate action at the national and international levels along with participation of the private sector to fund mitigation and adaptation projects in different sectors. For instance, the Global Facility for Disaster Reduction and Recovery (GFDRR), launched in 2006 and hosted by the World Bank with 21 participating donors and other stakeholders, offers a unique business model for advancing disaster risk management based on ex-ante support to high risk countries and ex-post assistance for accelerated recovery and risk reduction following a disaster (World Bank, 2013b). As of December 2012, GFDRR had provided support to over 80 countries and received US$ 342 million in pledges and contributions to implement its multi-annual work programme (World Bank, 2013b). Since 2008 the World Bank through the International Bank for Reconstruction and Development (IBRD) has been issuing Green Bonds to raise funds from investors in the capital markets to develop innovative solutions and attract private sector financing for climate action in IBRD borrowing countries, and raise awareness about climate change and the opportunities to invest in climate solutions (World Bank, 2013b). In India, the government has set up a National Adaptation Fund with initial funding of US$ 55.6 million to meet the adaptation needs of sectors such agriculture, water and forestry, in addition to sectoral spending by sector ministries (UNFCCC, 2015). India has also set up a National Clean Environment Fund to finance clean energy technologies and projects (UNFCCC, 2015). Such initiatives will enhance efforts to build a climate resilient future.

ABOUT THE BOOK

Keeping in view the above discussion, this book addresses the challenges and opportunities for building a climate resilient economy and society. The chapters included in this volume cover a broad range of topics such as vulnerability, adaptation and resilience, challenges and prospects for building climate resilience in different sectors such as agriculture, marine ecosystems, urban areas, energy and water-stressed areas, carbon pricing and financing, REDD+, climate policies and governance, among others. The chapters include those that have a global or regional focus as well as case studies drawn from a cross-section of countries in Africa, Asia, Europe and North America. The contributors to this book are leading experts from around the world who have made a significant contribution to the literature in this area. Some of them have also been part of IPCC Assessments. For convenience, the discussion in the book is organized in three parts: I, 'Vulnerability, adaptation and resilience'; II, 'Climate resilience: Sectoral perspectives'; and III, 'Incentives, governance and policy'.

Part I: Vulnerability, Adaptation and Resilience

Issues related to vulnerability, adaptation and resilience are the focus of the chapters included in Part I.

Encouraging adaptation is an essential aspect of the policy response to climate change, as noted earlier. However, given that human activities are the main cause of environmental transformations worldwide, it follows that adaptation itself also has the potential to generate further pressures, creating new threats for both local and global ecosystems. From this perspective, policies designed to encourage adaptation may conflict with regulations aimed at preserving or enhancing environmental quality. This aspect of adaptation has received relatively little consideration in either policy design or academic debate. To highlight this issue, Fezzi et al. in Chapter 2 analyse the trade-offs between two fundamental ecosystem services which will be impacted by climate change: provisioning services derived from agriculture and regulating services in the form of freshwater quality. Their results indicate that climate adaptation in the farming sector will generate fundamental changes in river water quality. In some areas, policies which encourage adaptation are expected to conflict with existing regulations aimed at improving freshwater ecosystems. These findings illustrate the importance of anticipating the wider impacts of human adaptation to climate change when designing environmental policies.

It is important to identify the factors influencing adaptive capacity among households within a community, as doing so will enable effective targeting of government interventions to address the risks posed by climate change. In Chapter 3 Panda et al. study such factors using household survey data collected from a drought-prone region of Orissa, India. In the survey respondents were asked about the adaptations that they had engaged in to deal with the risk of drought, as well as a number of indicators for adaptive capacity taken from the literature. The study found many indicators to correlate with one or more adaptations taken. However, many of these indicators, while increasing the likelihood that one adaptation would be taken, also decreased the likelihood that another would be taken, and hence were not unambiguous determinants of greater adaptive capacity in general. Access to crop insurance was found to be particularly effective: it correlated with an increased likelihood of engaging in two separate yield-raising adaptations. Their results suggest that further attention to crop insurance may be warranted, as well as further research to determine if the other indicators may be effective in other contextual settings.

Nature plays an important role in addressing the risks posed by climate change. In Chapter 4, Mutafoglu et al. explore nature's contribution to improving micro-climatic conditions in cities and mitigating urban heat stress, thereby helping cities become climate resilient. Green infrastructure, such as parks and tree-lined streets, can contribute to climate resilience and the health of urban populations by reducing heat stress, as well as hospitalizations and mortality. Today, with more than half of the global population urbanized, population densities and the heat island effect amplify heat-related risks in cities and necessitate appropriate solutions. The chapter presents a range of examples illustrating the benefits of nature, building mainly on insights from Europe. It also details how stakeholders collaborate to invest in urban and suburban green infrastructure and use a variety of tools, measures, processes and financing sources. The chapter then outlines a roadmap for moving forward.

In Chapter 5 Shandas et al. argue for using coproduction as a model for urban resilience, based on a recent collaborative project between planners and researchers. They review the findings of four collaborative workshops of neighbourhood-scale climate adaptation, using spatial and statistical analyses, as well as posit the conceptual framework underlying the project's collaborative rationality. Their assessment is from two perspectives: (1) an academic, phenomenological lens; and (2) a pragmatic lens. They argue that, in order to understand resilience, we must first ask the question 'Resilience to what and for whom?' To that end, they consider what characterizes those communities that are affected by two major urban environmental hazards, namely air pollution and urban heat, in the city of Portland, Oregon in the Pacific Northwest of the United States. Empirically, the authors explore neighbourhood level exposure to air pollution and extreme heat, as well as these spatially defined communities' physiological sensitivity to the two hazards, and their social capacity to adapt to them. Phenomenologically, the authors argue that this collaboration is an effort in *coproduction*, encapsulating cooperation between scientific experts and governmental authorities in the production of knowledge toward a socially determined goal in the public interest.

Small island developing states (SIDS) are most vulnerable to climate change. It is due to the insistence of SIDS that the Paris Climate Agreement concluded in December 2015 also set a desirable target of limiting temperature rise to 1.5 °C above pre-industrial levels. Chapter 6 by Pichs-Madruga discusses the economic, social and environmental challenges, and options for SIDS to address the risks posed by climate change, especially rising sea levels, extreme weather events and degradation of marine ecosystems with their consequent impacts on lives and livelihoods in the SIDS. Mitigation and adaptation measures, improving the sustainability of marine ecosystems and marine fisheries, disaster risk reduction and community-based approaches to development are critical to enhance climate resilience and sustainable development in SIDS.

Part II: Climate Resilience: Sectoral Perspectives

The risks posed by climate change differ across sectors, and hence policies to build climate resilience need to take account of these sectoral perspectives. In Part II the chapters focus on the challenges and opportunities for enhancing climate resilience in different sectors such as agriculture, marine ecosystems, urban areas, water-stressed regions and energy.

Agriculture is highly exposed to climate change, and much of the international climate discourse has focused on the sector's vulnerability, resilience and adaptive capacity for given projected climate scenarios. The sector also has a recognized role in terms of climate change mitigation. The agricultural sector accounts for approximately a third of total global anthropogenic GHG emissions, including emissions from land use change and deforestation (IPCC, 2014a). These climate challenges overlap a growing concern about global food security, which highlights additional stressors, including demographic changes, natural resource scarcity, and economic convergence in consumption preferences, particularly livestock products. The Food and Agriculture Organization (FAO) of the United Nations estimates that, to meet the demands of a larger population (expected to cross 9 billion by 2050), food supply will need to grow by 60 per cent. The term 'climate smart agriculture' (CSA) has been coined to describe practices, systems and institutional arrangements that balance the trade-offs inherent in meeting these objectives. Climate smart agriculture has become a focus of research in developed and developing countries. In Chapter 7, Arakelyan et al. provide a critical appraisal of the concept of CSA and review the experiences in implementing CSA in a cross-section of countries in Africa. The chapter then discusses the policies and institutional support needed to achieve climate smart agriculture as well as highlights some of the concerns regarding CSA, especially its sustainability and distributional consequences.

Rising sea levels, pollution, and increasing stresses on marine species and areas are major challenges confronting marine ecosystems as a result of climate change. In Chapter 8 Tisdell reviews the major changes in marine environments which natural scientists expect to occur because of elevated levels of GHGs in the atmosphere and considers their general implications for the presence of living organisms and supply of marine ecoservices. The chapter then focuses attention on the insights obtained from a Norwegian investigation of the economic impacts of GHG-induced changes in marine ecosystems. Thereafter GHG-induced losses of coral reefs and their potential economic impacts are discussed. Proposed policies for responding to predicted changes in marine ecosystems are examined, followed by a discussion of building resilience in the context of marine ecosystems.

Sumaila et al. in Chapter 9 review current knowledge on the potential impacts of climate change on marine ecosystems and the millions of people worldwide who depend on them. The authors highlight the fact that different parts of the world would be impacted differently and, therefore, climate will impact people in different regions differently. They then provide a number of policy recommendations to help prepare society for the changes that we are already seeing and those yet to come. To build climate resilient marine ecosystems and global fisheries, the authors highlight the need to transform their management by increasing incentives for community engagement and deploying marine protected areas, and to promote sustainability enhancing public policies by avoiding harmful ones such as the provision of capacity enhancing fisheries subsidies.

Climate change is projected to aggravate water stress in dry and arid regions. Against the backdrop of the epic drought faced by California in the USA in 2015, Chapter 10 by Miller discusses the significant vulnerabilities and options for maintaining the resilience of the state's water-dependent economic activities. The drought led to very uneven impacts on different water users and sections of the state, as well as on natural ecosystems versus managed landscapes. Differential vulnerabilities can be traced to the state's complex geography, the configuration of its water storage and delivery infrastructure, and its imperfectly administered mixture of prior appropriation and riparian surface water rights coupled with limited regulation of groundwater withdrawals. Approaches for reducing economic losses have included selective fallowing, increased groundwater pumping, adoption of water-saving irrigation techniques, and market transfers of water. The chapter highlights innovative water management strategies that have emerged over the course of the drought and the lessons that California's drought experience suggests for other areas that may face increasing drought risks in a warmer future climate.

Urban areas will face increased risks as a result of climate change such as heat stress, storms and extreme precipitation, inland and coastal flooding, air pollution, water scarcity, sea level rise, storm surges, and so on (IPCC, 2014a). By the year 2050 about 66 per cent of the world's population will reside in urban areas; much of the projected increase will take place in Asia and Africa (United Nations, 2014). Recent studies confirm that the impacts of climate change are already being seen and felt in major Asian cities. To combat the impacts of climate change, Asian city governance needs to mainstream climate change resilient policies into urban planning. Chapter 11 by Johnson et al. explores the challenge of enhancing climate resilience in two of the world's most vulnerable coastal city regions – Dhaka (Bangladesh) and Ho Chi Minh City (Vietnam). The chapter reviews the policies and governance structures of these cities, comparing the extent to which national and municipal governments could implement measures aimed at building climate resilience at the urban scale.

Fossil fuel-led growth has been a major contributory factor to rapid rise in carbon emissions. Emissions of CO_2 from fossil fuel combustion and industrial processes contributed about 78 per cent of the total GHG emissions increase from 1970 to 2010 (IPCC, 2014a). Cutting down on use of fossil fuels and shifting to clean energy sources is therefore a major strategy for combating global warming. Chapter 12 by Timmons discusses the economic principles that should govern renewable energy choices. Renewable energy sources including biomass energy, water power, wind, solar power and geothermal energy have somewhat different characteristics from fossil fuels: they are capital intensive, with their costs dependent on interest rates; their costs are highly dependent on their scales and production sites; and many renewable energy sources are available only intermittently. Minimizing total cost of providing renewable energy suggests that marginal costs of individual renewable energy sources be equal. In many areas, use of more expensive sources such as solar photovoltaic energy will thus make it economical to develop hydropower and wind power on sites that might not appear feasible currently. Similarly, the marginal cost of renewable energy suggests that additional energy conservation will be economical, and a large portion of the transition to renewable energy will likely be accomplished through energy conservation rather than energy production. To minimize total costs, equality of marginal costs must also hold at all points of time and from all points in space, suggesting possibilities for energy storage and long-distance energy transmission facilities. While the market would eventually accomplish a renewable energy transition owing to rising fossil fuel prices, public policy will likely be needed to make the renewable energy transition soon enough to avoid the worst effects of climate change.

Part III: Incentives, Governance and Policy

Building climate resilience requires an enabling environment in terms of appropriate incentives, governance, institutions and policies. These issues are examined in the chapters included in Part III.

Over the past 25 years, a growing number of countries have adopted policies that place a price on the emissions of carbon dioxide (CO_2) and other GHGs. Regardless of whether such carbon pricing is implemented through taxes or emissions trading schemes (ETS), these policies can raise substantial amounts of public funds. How should the revenues raised by carbon pricing policies be managed? Chapter 13 by Barrage surveys both the economic models and the policy approaches that have been used to address this question from the perspectives of economic efficiency, equity and political feasibility. First, from an efficiency-maximizing perspective, public finance models typically find that capital income tax reductions would be the optimal use of carbon revenues. Recycling the revenues through labour income tax cuts ranks second, whereas rebating the revenues directly to households as lump-sum payments entails the largest efficiency costs. In contrast, based on equity considerations, lump-sum rebates can be preferable to labour tax reductions in addressing the regressivity of carbon prices. In contrast, an ETS with free permits and corporate income tax cuts is likely the most regressive option, but would compensate industries and thus achieve political feasibility at the lowest cost.

Reducing emissions from deforestation and forest degradation (REDD+) is a recent global response to the challenges posed by emissions from deforestation and forest degradation. Global interest in REDD+ is growing and warrants a comprehensive review of what REDD+ is, how it is implemented and what key issues need to be considered to steer REDD+ towards a climate resilient socio-ecological system. Chapter 14 by Mohammed and Inoue indicates that REDD+ can be understood as global multilevel forest governance that links different actors at different scales, from local to global (glocal). REDD+ has evolved from the narrowly scoped reducing emissions from deforestation (RED) to the current REDD+, which has considered not only deforestation but also forest degradation, sustainable forest management, and social and biodiversity safeguards through repeated engagement and negotiation among glocal actors since the 2005 Cancun IPCC meeting. Currently, a total of eight actors, ranging from multilateral financial institutions to national governments and indigenous people, are identified as playing an important role in the negotiation and/or design and/or implementation of REDD+. The authors identify five salient issues that need to be addressed to achieve a resilient socio-ecological system. These are: lack of a strong global institution; biodiversity concerns; lack of competitiveness; lack of tenure clarity; and prospects for distributive and procedural injustice.

Giving favoured access to climate finance and clean technologies is critical to enabling developing countries to transit to a low carbon economy and society. Global climate finance to developing countries is set to rise with the establishment of the Green Climate Fund. To be effective, climate finance must reach and be prioritized by the communities that need it most and be used to fund solutions that work on the ground. To achieve this, mechanisms need to be put in place to channel the money from national level to local communities in a way that is transparent, participatory and efficient. The institutional architecture of existing devolved or decentralized government provides a ready-made framework which offers good value for money and will be sustainable as finance flows increase in the future. Orindi et al.'s analysis (Chapter 15) of the Adaptation Consortium in Kenya indicates that devolved County Climate Change Funds are proving to be an effective mechanism to deliver climate finance in support of community-prioritized investments in public goods that build local resilience to climate change.

The European Union (EU) has a consolidated climate and energy policy, which has played a pioneering role by adopting a wide range of emission reduction measures. However, it is often claimed that these measures have a negative effect on the economy, especially in terms of growth and competitiveness. Chapter 16 by Carraro and Davide reviews the recent literature on the European experience to understand if these concerns are true. Their analysis primarily focuses on studies assessing major economic indicators, such as costs, competitiveness, carbon leakage and income distribution, with the objective of highlighting both the limits and the opportunities of the EU's regulatory framework, as well as its potential for reconciliation with socio-economic objectives.

The atmosphere is a community asset that belongs to all people. The problem is that it is treated as an open access resource: anyone can emit carbon dioxide into the atmosphere with no consequences to themselves but huge cumulative consequences to the climate and the global community. Many agree that charging companies and individuals for the damages their emissions cause, for example a comprehensive carbon tax or cap/auction/dividend/trade system, would drastically cut emissions. However, despite some interesting regional experiments, implementing this kind of system via international negotiations at the global scale has proven close to impossible. A few critical governments, influenced too much by fossil fuel interests, have been blocking binding commitments and effective economic instruments. In Chapter 17, Costanza argues that global civil society can change this if it *claims property rights* over the atmosphere. By asserting that all people collectively own the sky, legal institutions surrounding property can be used to protect our collective rights, charge for damages to the asset and provide rewards for improving the asset. This idea has been proposed by Peter Barnes and others. The public trust doctrine is a powerful emerging legal principle that supports this idea. The doctrine holds that certain natural resources are to be held in trust as assets to serve the public good. It is the government's responsibility as trustee to protect these assets from harm and maintain them for the public's use. Under this doctrine, the government cannot give away or sell off these public assets to private parties. The public trust doctrine has been used in many countries in the past to protect water bodies, shorelines, fresh water, wildlife and other resources.

REFERENCES

Adgers, W.N. (1999), 'Social vulnerability to climate change and extremes in coastal Vietnam', *World Development*, **27** (2), 249–269.

Ahmed, M. and S. Suphachalasai (2014), *Assessing the Costs of Climate Change and Adaptation in South Asia*, Manila: Asian Development Bank.

Cammack, D. (2007), 'Understanding the political economy of climate change is vital to tackling it', *Opinion* (Overseas Development Institute), **92**.

Dayal, A. (2014), 'The need for climate change resilience', Rockefeller Foundation, 27 August, https://www.rockefellerfoundation.org/blog/need-climate-change-resilience/ (accessed 21 September 2016).

EJF (2006), *Mangroves: Nature's Defence against Tsunamis: A Report on the Impact of Mangrove Development and Shrimp Farm Development on Coastal Defences*, London: Environmental Justice Foundation, http://ejfoundation.org/sites/default/files/public/tsunami_report.pdf.PdfCompressor-1022348.pdf.

Folke, C. (2006), 'Resilience: The emergence of a perspective for social-ecological systems analyses', *Global Environmental Change*, **16**, 253–267.

IPCC (2007), *Climate Change 2007 Synthesis Report Summary for Policymakers*, Geneva: IPCC, WMO and UNEP, https://www.ipcc.ch/publications_and_data/ar4/syr/en/spms3.html (accessed 10 October 2016).

IPCC (2014a), *Climate Change 2014: Synthesis Report Summary for Policymakers*, Geneva: IPCC, WMO and UNEP, https://www.ipcc.ch/pdf/assessment-report/ar5/syr/AR5_SYR_FINAL_SPM.pdf (accessed 18 September 2016).

IPCC (2014b), 'Summary for policymakers', in C.B. Field, V.R. Barros, D.J. Dokken,

K.J. Mach, M.D. Mastrandrea, T.E. Bilir, M. Chatterjee, K.L. Ebi et al. (eds), *Climate Change 2014: Impacts, Adaptation, and Vulnerability, Part A: Global and Sectoral Aspects: Contribution of Working Group II to the Fifth Assessment Report of the Intergovernmental Panel on Climate Change*, Cambridge, UK: Cambridge University Press, pp. 1–32.

Munich RE (2015), 'Loss events worldwide 1980–2014', NatCatSERVICE, Munich RE, January, https://www.munichre.com/site/touch-naturalhazards/get/documents_E2080665585/mr/assetpool.shared/Documents/5_Touch/_NatCatService/Focus_analyses/1980-2014-Loss-events-worldwide.pdf (accessed 27 September 2016).

Nelson, D.R., W.N. Adger and K. Brown (2007), 'Adaptation to environmental change: Contributions of a resilience framework', *Annual Review of Environment and Resources*, **32**, 395–419.

Ninan, K.N. (2016), 'Climate change: Economics, environment and public policy (with special reference to India)', ICCR public lecture delivered at the Department of Economics, University of Gothenburg, Sweden, 30 May.

NOAA (2016), '2015 state of the climate: Carbon dioxide', https://www.climate.gov/news-features/featured-images/2015-state-climate-carbon-dioxide (accessed 26 September 2016).

Olivier, J.G.L., G. Janssens-Maenhout, M. Muntean and J.A.H.W. Peters (2015), *Trends in Global CO_2 Emissions 2015 Report*, The Hague: PBL Netherlands Environmental Assessment Agency and Ispra: European Commission Joint Research Centre.

Panda, A. (2016), 'Vulnerability to climate variability and drought among small and marginal farmers: A case study in Odisha, India', *Climate and Development*, published online 16 June, http://dx.doi.org/10.1080/17565529.2016.1184606.

Royal Society (2014), *Resilience to Extreme Weather*, Royal Society Science Policy Report 02/14, www.royalsociety.org/resilience (accessed 17 September 2016).

SCBD (2009), *Connecting Biodiversity and Climate Change Mitigation and Adaptation: Report of the Second Ad Hoc Technical Expert Group on Biodiversity and Climate Change*, CBD Technical Report 41, Montreal, Canada: Secretariat of the Convention on Biological Diversity and UNEP.

Stern, N. (2007), *The Economics of Climate Change*, Stern Review, Cambridge, UK: Cambridge University Press.

Tol, R.S.J. (2009), 'The economic effects of climate change', *Journal of Economic Perspectives*, **23** (1) (Spring), 29–51.

Tol, R.S.J. and G.W. Yohe (2006), 'A review of the Stern Review', *World Economics*, **17** (4), 233–250.

UNEP (2015), *The Emissions Gap Report 2015: A UNEP Synthesis Report*, November, http://uneplive.unep.org/media/docs/theme/13/EGR_2015_301115_lores.pdf (accessed 28 September 2016).

UNFCCC (2015), 'India's intended nationally determined contribution: Working towards climate justice', http://www4.unfccc.int/submissions/INDC/Published%20Documents/India/1/INDIA%20INDC%20TO%20UNFCCC.pdf (accessed 28 September 2016).

UNFCCC (2016), 'Paris Agreement: Status of ratification', http://unfccc.int/paris_agreement/items/9485.php (accessed 11 October 2016).

United Nations (2014), *World Urbanisation Prospects 2014 Revision: Highlights*, New York: United Nations Department of Social and Economic Affairs, https://

esa.un.org/unpd/wup/Publications/Files/WUP2014-Highlights.pdf (accessed 15 October 2016).
World Bank (2013a), *World Development Report 2014: Risk and Opportunity: Managing Risk for Development*, Washington, DC: World Bank.
World Bank (2013b), *Building Resilience: Integrating Climate and Disaster Risk into Development: Lessons from the World Bank Group Experience*, Washington, DC: World Bank.

PART I

Vulnerability, adaptation and resilience

2. The environmental impact of climate change adaptation on land use and water quality

Carlo Fezzi, Amii R. Harwood, Andrew A. Lovett and Ian J. Bateman

INTRODUCTION

Encouraging adaptation is an essential aspect of the policy response to climate change (Pielke et al., 2007). Adaptation seeks to reduce the harmful consequences and harness any beneficial opportunities arising from the changing climate. However, given that human activities are the main cause of environmental transformations worldwide (Vitousek et al., 1997), it follows that adaptation itself also has the potential to generate further pressures, creating new threats for both local and global ecosystems. From this perspective, policies designed to encourage adaptation may conflict with regulation aimed at preserving or enhancing environmental quality. This aspect of adaptation has received relatively little consideration in either policy design or academic debate. To highlight this issue, this chapter analyses the trade-offs between two fundamental ecosystem services which will be impacted by climate change: provisioning services derived from agriculture and regulating services in the form of freshwater quality. Results indicate that climate adaptation in the farming sector will generate fundamental changes in river water quality. In some areas, policies which encourage adaptation are expected to be in conflict with existing regulations aimed at improving freshwater ecosystems. These findings illustrate the importance of anticipating the wider impacts of human adaptation to climate change when designing environmental policies.

ANALYSIS

On a global scale, agriculture is the economic sector which is likely to bear the greatest financial impact as a result of climate change (Lobell et al.,

2011). Farmers are expected to adapt by switching activities to those which are most profitable given the new conditions they will face. Since agriculture is one of the major drivers of freshwater quality (Vitousek et al., 1997; Sterling et al., 2013), these changes in farmland use have the potential to substantially alter water ecosystems. For example, agricultural inputs are responsible for nutrient overload and eutrophication in water bodies worldwide (Vitousek et al., 1997; McIsaac et al., 2001; Conley, 2012) and are a major focus of policy action, e.g. US Clear Water Act (EPA, 2012), EU Water Framework Directive (European Commission, 2000). Understanding the impact of agricultural adaptation to climate change on water quality is, therefore, essential for delivering harmonized and efficient policies (although, from a theoretical standpoint, if all the external effects of agriculture on the environment were correctly priced, i.e. internalized, the market would automatically deliver socially optimal outcomes).

An important feature of the relationship between farming and water quality is its strong spatial heterogeneity. Agricultural activities, adaptation options and environmental quality vary significantly over relatively small areas. Therefore, a meaningful analysis requires data reflecting this fine-scale variation, which would be irremediably overlooked if large-scale, aggregated data were employed (Willis and Bhagwat, 2009; Goldstein et al., 2012). Our empirical investigation focuses on Great Britain (GB), where detailed and long established information sources allowed us to assemble a unique dataset, spanning more than 40 years at a resolution of 2km grid squares (400 ha). This constitutes about half a million spatially referenced, time-specific, land use records (see methods and supplementary materials, SM, sections S1.2 and S2.2 in Fezzi et al., 2015). Almost 80% of GB's land use is devoted to a very heterogeneous farming system, ranging from the intensive arable cropping of the English lowlands to the extensive grazing farms of the upland northern and western regions, including much of Scotland and Wales. While water quality in GB freshwater bodies is subject to several EU Directives (European Commission, 1991, 2000), a large share of its rivers and lakes are still characterized by high nutrient concentrations which fail to comply with existing regulations.

Our analysis is based on an integrated framework linking a spatially-explicit econometric model of agricultural production to a statistical model of river water quality. Integrating economic models of land use change with environmental models predicting consequent impacts on multiple ecosystem services has been a focus of considerable recent research (Goldstein et al., 2012; Bateman et al., 2013; Johnson et al., 2014; Lawler et al., 2014; Polasky et al., 2014). By integrating novel land use and water quality models, our analysis examines how adaptation to climate change in agriculture is expected to affect aquatic ecosystems. By examining how spatial heterogeneity in climate has influenced agricultural production decisions and farm income (farm gross margin, FGM) (Fezzi and Bateman, 2011; Bateman et al., 2013) to date, we project how farmers will adapt to future climate. To estimate resulting water quality impacts, we rely on spatially explicit statistical models linking land use to observed concentrations of nitrate (NO_3) and phosphate (as phosphorus, P) in rivers.

Our agricultural production model builds upon a strand of research in agricultural economics (Fezzi and Bateman, 2011; Kling, 2011). We develop a structural econometric model with a flexible specification of the effects of climate on agricultural land use and production (see SM S1.3 in Fezzi et al., 2015). Temperature and precipitation are represented via linear regression splines coupled with a fixed effect estimator to both control for unobserved missing variables and isolate the impact of climate. Even within the relatively small area of GB, variation in climatic and environmental conditions is sufficient to yield substantial differences in agricultural productivity and, hence, land use. These differences are captured by the model along with variation due to other drivers such as changes in policies and prices.

Figure 2.1 reports the estimated impact of temperature and precipitation on two illustrative land use shares (arable and temporary grassland) and on beef cattle rates (heads/ha). As shown in the upper row, arable is the dominant land use in low precipitation areas, with pastures becoming more common only as rainfall rises. Beef cattle stocking rates rise rapidly with precipitation (and the concomitant increase in pasture size) until rainfall reaches about 500mm, after which cattle rates begin to slowly decline as they are replaced by more resilient livestock such as sheep. Considering the effect of temperature, in the second row, we observe a positive relationship with the share of arable land, related to the effect on yield which, however, becomes gradually less steep and finally negative for the highest temperatures, confirming previous research findings (Fezzi and Bateman, 2011; Lobell et al., 2011; Bateman et al., 2013).

We analyse water quality via statistical models explaining observed river nitrate and phosphate concentrations as functions of the land use and the climate characterizing the land upstream from each water monitoring station, derived via the use of a geographical information system (GIS) (see 'Methods'). By including fixed effects, we estimate coefficients which are robust to potential un-observed confounders. The parameters of the final models are reported in Table 2.1.

The land share coefficients should be interpreted relative to the omitted land use category, which here is arable farming. Therefore, negative (positive) coefficients indicate that a land use produces less (more) pollution than arable. All parameters conform to our expectations and

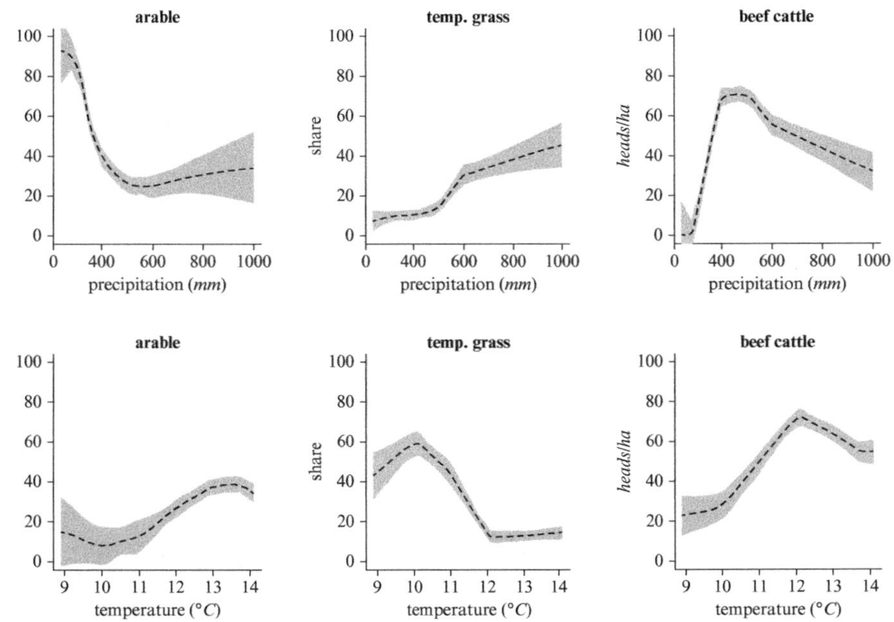

Notes:
Total precipitation in mm; average temperature in °C; land use shares as percentage of agricultural area; beef cattle stocking rates in heads/ha. Dashed lines indicate estimated relations, grey areas the 95% asymptotic confidence intervals. All other explanatory variables are fixed at the sample.

Figure 2.1 Estimated impact of total precipitation and average temperature during the growing season (April–September) on land use shares and beef cattle stocking rates

Table 2.1 Water quality models

	Nitrate	Phosphate
Intercept	46.48***	0.389***
	(2.57)	(0.056)
$share_{urban}$	−4.24	0.897***
	(20.08)	(0.137)
$share_{rough}$	−40.23***	−0.485*
	(7.43)	(0.246)
$share_{grass}$	−37.94***	−0.311***
	(9.47)	(0.132)
$share_{wood}$	−34.64***	−0.589*
	(9.31)	(0.339)
$D_{livestock}$ * $share_{grass}$	10.38**	–
	(4.93)	
D_{pop} * $share_{urban}$	0.18	–
	(0.53)	
precipitation	−0.62**	–
	(0.92)	
σ	7.47***	0.231***
	(0.39)	(0.011)
Log-likelihood	−286.26	−439.60
Pseudo R^2 (McFadden)	0.28	0.10

Notes:
Interval regression model estimated with Gaussian residuals on 214 monitoring points located on independent river catchments.
Coefficients need to be interpreted using the share of arable land as the baseline category.
$D_{livestock}$ is the livestock density (number of cattle per hectare of grassland); D_{pop} is the population intensity (defined as the number of people per hectare).
Significance levels: * = 0.10, ** = 0.05, *** = 0.01.

previous literature (McIsaac et al., 2001; Conley, 2012; Sterling et al., 2013). Considering nitrate, urban land yields levels of concentration which are not significantly different from those of arable, while other land uses generate lower leaching. In the extreme, an entirely arable catchment is predicted to generate average nitrate concentrations of just over 44mg NO_3/l, which would be considerably above the threshold of 30mg NO_3/l identified by EU regulations (European Commission, 1991, 2000).

Similar consistency with previous research (Jarvie et al., 2006) is confirmed within the model of phosphate. The estimates indicate that the main source of phosphorus in rivers is urban land, which has a coefficient almost three times higher than that of arable, again represented by the intercept. Nevertheless, the model suggests that a river catchment draining an entirely arable area would typically yield a concentration of about 0.39mg P/l, or above the threshold of 0.2mg P/l recommended by the WFD (European Commission, 2000), while a fully urbanized catchment is predicted to yield concentrations averaging around 1.29mg P/l. Again, less intensive land uses produce significantly lower concentrations.

We integrate the agricultural land use and river quality models and verify their performance in predicting observed data via out-of-sample testing (see SM S3 in Fezzi et al., 2015). In order to project the impact of climate change adaptation, we hold prices, policy and technological change constant at their baseline values. In addition, we also leave unchanged all non-agricultural land allocation and farm woodland, which is mainly driven by area-specific governmental and planning policies. Therefore, these scenarios are not projections of the future, but rather illustrate, *ceteris paribus*, the impact of climate change adaptation. In GB climate change is expected to generate a warmer and drier growing season, with average temperatures projected to increase by about 2°C, and total precipitation to decrease, on average, by about 60mm, by the 2040s (UKCIP, 2011) (see also SM, figure S3, Fezzi et al., 2015).

Figure 2.2 summarizes our findings for the baseline year and climate change scenarios in the 2020s and 2040s. The first column of maps shows baseline conditions for agricultural production values (map A), concentrations of nitrate (map D) and phosphate (map G). Current agricultural production shows a clear South–North divide, with the lowlands in the south being significantly more profitable than the colder and wetter regions in Scotland and Wales. Our results indicate that climate change will reduce this gap, primarily benefiting northern regions as higher temperatures will allow increases in more profitable arable and higher livestock intensity (maps B and C). However, such changes are also expected to amplify the pressure on the environment, increasing diffuse emissions into rivers. Overall, the area of land at risk of reporting high nitrate and high phosphate concentrations is projected to increase by 30% (1.4 million ha) and 20% (1.6 million ha) respectively, as a result of climate change adaptation (see SM S4.3 in Fezzi et al., 2015). These areas are illustrated in black in maps E, F, H and I. This indicates that adaptation will significantly increase the effort required to achieve water quality standards, particularly in the eastern uplands and midlands where temperature rises will permit significant increases in agricultural production.

Map A in Figure 2.3 summarizes the spatially heterogeneous effects of climate change adaptation on agricultural incomes and water quality by the 2040s. Areas where adaptation to climate change will yield improvements in farming without significant environmental repercussions are shown in medium grey (including most of the North-West but the highest upland regions). Other areas, which are not expected to yield reductions

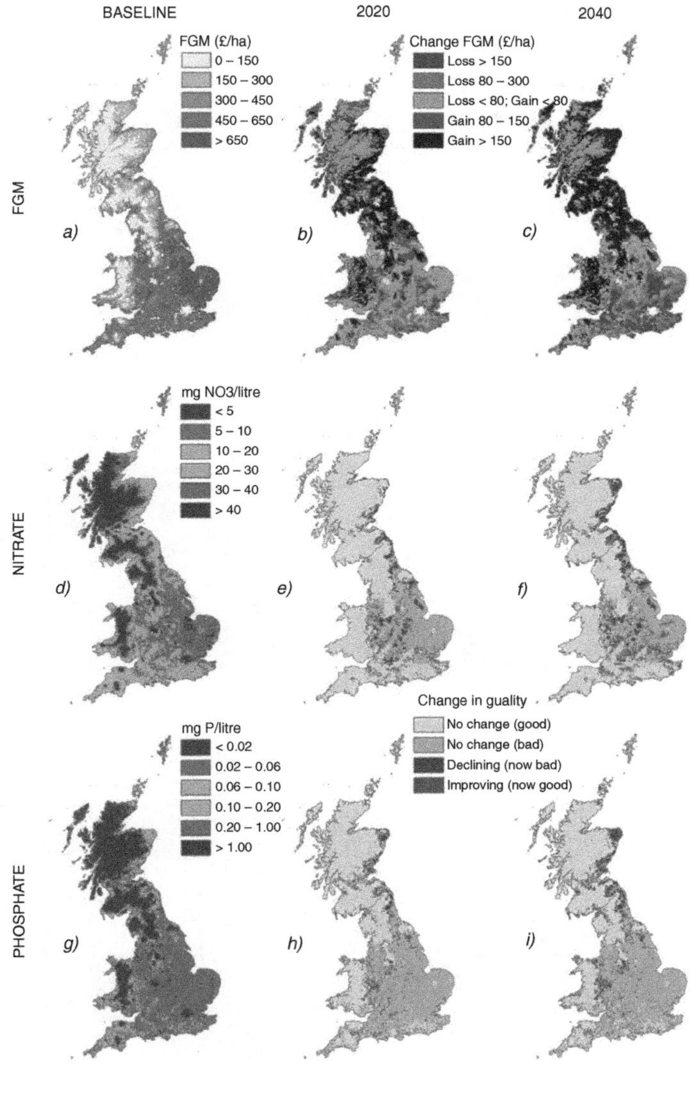

Notes:
Climate change is UKCP09 medium emission, SRES A1 B scenario; farm gross margin in £/ha; river quality in NO_3, P.
The first column (maps A, D, G) presents estimated values for the baseline climate. The second (maps B, E, H) and third (maps C, F, I) columns present projected changes for the 2020s and 2040s (UKCIP medium emission scenarios).
The 2020s and 2040s are defined respectively as the climate averages for the years 2010–2039 and 2030–2059 as by the UKCIP (Nakicenovic and Swart, 2000).

Figure 2.2 Impact of climate change for the 2020s and 2040s on farm gross margin and river quality

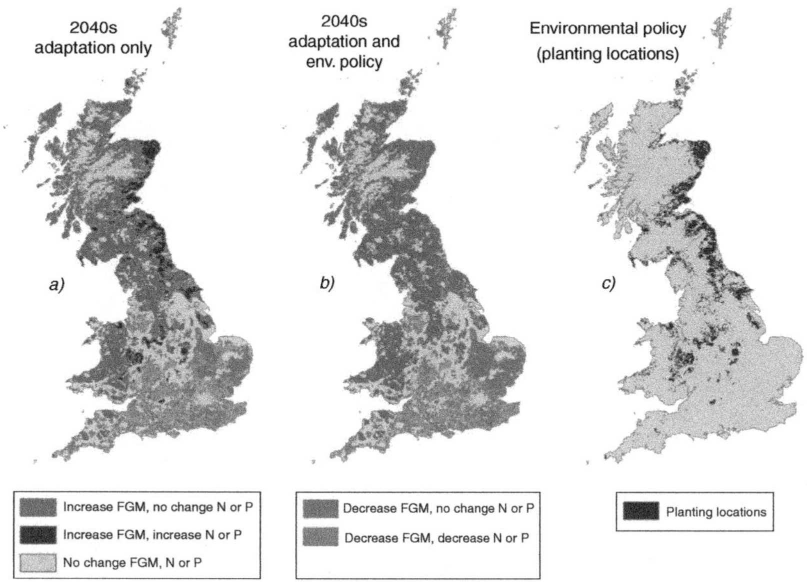

Notes:
Map A shows areas of significant increase (decrease) in profit defined as FGM change > (<) 80 £/ha (see Figure 2.2), and water quality with decreases defined as cases where N or P concentrations are low or moderate in the baseline (i.e. lower than 30 mg NO3/l and 0.2 mg P/l respectively; see SM S2 in Fezzi et al., 2015) and become high in the climate change and agricultural adaptation scenario (the converse applies for the definition of water quality improvement). Map B illustrates these various changes following the introduction of a policy response consisting of the planting of 500 000 hectares of new broadleaf forests in the locations indicated in Map C.
The 2040s period is defined as the climate averages for the years 2030–2059 as given by UKCIP (2011).

Figure 2.3 The impact of climate change adaptation and possible policy response

in water quality but are predicted to see falls in farm income are shown in light grey (principally in the south of England). The map also reveals areas of trade-off, either regions where adaptation is expected to raise FGM at the expense of generating high nutrient concentrations (shown in dark grey in areas such as the North-Eastern coast and across parts of the English midlands), or where losses in farm income will be accompanied by improvements in water quality (medium-dark areas in the south). Remaining areas are not expected to experience substantial changes in either farm incomes or diffuse pollution.

In considering a potential policy response to the problem of adaptation-induced deterioration of river water quality, an option within the British context is provided by recent government announcements regarding an intention to significantly extend woodland coverage over the next decades (Defra, 2014; Natural Capital Committee, 2014). Among the diverse set of benefits which can be generated by forests (including carbon storage, recreational provision, timber output, etc.), this initiative also views water quality enhancements as a key argument for woodland creation, given the very low nutrient leaching rates generated by this land use. Therefore, we examine the effect of locating the woodland in those areas where adaptation is expected to generate the largest falls in water quality. Map C in Figure 2.3 shows planting locations for 500 000ha of new forests (a level consistent with policy discussions) (Defra, 2014; Natural Capital Committee, 2014) while central Map B reveals the environmental and economic impacts of such a policy (results for different planting acreages are given in SM, table S8 in Fezzi et al., 2015). The effects are very significant, with almost all rivers in the targeted areas projected to remain in good condition despite the increase in agricultural production. This demonstrates how a systemic approach to interventions can anticipate the environmental impacts of climate change adaptation and deliver more than one policy-goal at the same time.

CONCLUDING REMARKS

As our discussion suggests, the potential effects of adaptation in the farming sector are not restricted to water quality. Adaptation may impact on water availability, wildlife, biodiversity, carbon sequestration, recreation, etc. On the other hand, climate change could also reduce the viability of agriculture in some areas, potentially diminishing certain pressures. Furthermore, the environmental impacts of adaptation are not limited to farming, but concern most activities which will be impacted by climate change, including energy demand and production (De Cian et al., 2013), fisheries (Barange et al., 2014), forestry (Guo and Costello, 2013) and health (Patz et al., 2005). This of course does not imply that adaptation is inappropriate; rather it demonstrates that policies should take into account the wider implications of adaptation and seek to incorporate such synergies and trade-offs. This will require a degree of integration across policy fields which is still lacking in current decision making (Nisson et al., 2012).

Methods

Land use model

The large database used for estimating the agricultural land use model was assembled using a variety of spatially explicit information. Land use and livestock data were derived from the June Agricultural Census (source, EDINA, www.edina.ac.uk). Collected on a 2km grid square (400ha) basis, this covers the entirety of GB for ten unevenly spaced years from 1972 to 2004. This constitutes roughly 55000 grid-square records per year, amounting to over 500000 grid-square observations for the overall analysis. We consider four categories of land use, each associated with different levels of pollution: (a) temporary grassland, (b) permanent grassland, (c) rough grazing and (d) arable (definitions in SM 1.2 in Fezzi et al., 2015). We include three livestock types: dairy cattle, beef cattle and sheep. Environmental drivers of agricultural land use include average temperature and accumulated rainfall, environmental and topographic variables, policies, etc. Yearly and regional fixed effects allow us to control for time- and spatially-varying omitted factors (see SM 1.2 in Fezzi et al., 2015).

We assume that farmers choose their land use activities (l_h) by taking into account expected input (**p**) and output (**w**) prices, policy constraints, climate and land quality (all included in the vector **z**). The agricultural land within each 400ha cell is modelled as an individual farm characterized by a multiproduct profit (π) function, which is maximized according to the following objective function:

$$\pi(\mathbf{p},\mathbf{w},\mathbf{z},L) = \max_{l_1,\ldots,l_h}\left\{\pi(\mathbf{p},\mathbf{w},\mathbf{z},,l_1,\ldots,l_h):\sum\nolimits_{i=1}^{h} l_h = L\right\}.$$

Using a normalized quadratic empirical specification for π^L and applying Hotelling's lemma, we derive land use share equations and land use intensity equations in linear forms (Fezzi and Bateman, 2011; Bateman et al., 2013) (SM S1.3 in Fezzi et al., 2015). For instance, if p_i indicates the price of cereals, the equation corresponding to cereal yield y_i is:

$$\frac{\partial\pi}{\partial p} = y_i = k_i + \mathbf{z}'\alpha_i + \mathbf{p}'\beta_i + \mathbf{w}'\gamma_i,$$

where k_i, α_i, β_i, γ_i are the parameters of the cereal yield equation to be estimated. As our data contains corner solutions (not all farms cultivate all possible crops), adding Gaussian disturbances and implementing ordinary least squares or generalized least square estimation leads to inconsistent results. Therefore, we implement a quasi maximum likelihood, heteroskedastic, simultaneous equation, Tobit model (Amemiya, 1973; Bateman et al., 2013). Predictive performance is tested via a rigorous out-of-sample forecasting exercise (see SM 1.3, 1.4, table S1, figure S1 in Fezzi et al., 2015).

Water quality model

Data on nitrate and phosphate concentration are extracted for over 5000 monitoring points collected as part of the General Quality Assessment (GQA) survey conducted annually by the Environment Agency to monitor the state of GB freshwater ecosystems (EA-AfA163, 2012). We selected data averages for the years 2005 to 2007 to fall within the period of our land cover and land use intensity information (see below and SM 2.2 in Fezzi et al., 2015). Since monitoring points can refer to stations located on the same river, or to rivers belonging to the same catchment, nutrient concentrations can be spatially dependent across stations. To implement standard statistical modelling on a sample of independent observations, we selected a smaller sub-sample of 214 stations belonging to non-overlapping catchments representing the locations and the range of nitrate levels observed in the full sample (see figure S2 and table S2 in Fezzi et al., 2015). GQA data classify nutrient levels as belonging to one of six categories from very low concentrations of pollution (highest water quality) to very high levels (worst quality), as detailed in SM 2.2 in Fezzi et al. (2015). Given the structure of this data, we model concentrations for nutrient q (nitrate or phosphate) at point j using interval regression techniques, which are generalizations of the censored Tobit model (Amemiya, 1973), as follows:

$$N_{q,j} = \mathbf{x}'_{jq}\mathbf{b}_q + e_{jq},$$

where $\mathbf{x}_{jq}$ indicates the matrix of explanatory variables, e_{jq} indicates an identically distributed residual term and $\mathbf{b}_q$ is the vector of parameters to be estimated. As explanatory variables we consider land use (arable, improved grassland, rough grassland, forest and urban), livestock intensity and population upstream from each GQA monitoring point, derived by weighted flow accumulation techniques (Kennedy, 2009) (see SM 2.2 in Fezzi et al., 2015). We include regional fixed effects to account for spatial omitted variables. Different model specifications with corresponding goodness of fit measures are reported in SM, table S3 in Fezzi et al. (2015).

Integrated framework

The land use model and the water quality model are estimated using the same spatial units and variable definitions. This ensures that a full integration of the two models is relatively straightforward. This integrated framework is verified via out-of-sample predictions (see SM 3, table S4 and figure S3 in Fezzi et al., 2015).

Climate change scenarios

We consider medium emission (Nakicenovic and Swart, 2000), climate change scenarios published by UKCIP (2011) as 25km grid-square projections for the '2020s' (defined as the average climate in years 2010–2039) and '2040s' (2030–2059) periods. Consistent with UKCIP, we use as a baseline the climate averages for the years 1961–1990 (see SM 4 in Fezzi et al., 2015). Table S5 in Fezzi et al. (2015) provides descriptive statistics of the climatic variables in the historical baseline and in each scenario, which are also represented via maps in figure S4 in Fezzi et al. (2015). Table S6 (see SM in Fezzi et al., 2015) provides descriptive statistics of our land use projections; table S7 (Fezzi et al., 2015) reports projection of nutrients' concentrations.

ACKNOWLEDGEMENTS

This chapter was originally published as Fezzi et al., (2015). The authors are grateful to the publishers for permission to reproduce this paper within the present volume. This research was supported by the LUCES Project (Project ID: 302290), funded by the European Commission under the Marie Curie International Outgoing Fellowship Programme, the SEER Project, funded by the ESRC (ref: RES-060-25-0063) and WEPGN, Brock University. Many thanks to Richard Carson, Silvia Ferrini and Kurt Schwabe for their helpful comments on a previous version of this chapter. Thanks to Anthony De-Gol for his assistance with the UKCIP data.

The analysis was designed by Carlo Fezzi with contributions from Ian J. Bateman and all the authors; Amii R. Harwood and Andrew A. Lovett undertook the data collection and the GIS analysis; Carlo Fezzi undertook the econometric analysis of the land use and the water quality models; Carlo Fezzi and Ian J. Bateman wrote the chapter with contributions from all the authors. The authors declare no competing financial interests.

REFERENCES

Amemiya, T. (1973), 'Regression analysis when the dependent variable is truncated normal', *Econometrica*, **41**, 997–1016.

Barange, M., G. Merino, J.L. Blanchard, J. Scholtens, J. Harle, E.H. Allison, J.I. Allen, J. Holt et al. (2014), 'Impacts of climate change on marine ecosystem production in societies dependent on fisheries', *Nature Climate Change*, **4**, 211–216.

Bateman, I.J., A. Harwood, G.M. Mace, R.T. Watson, D.J. Abson, B. Andrews, A. Binner, A. Crowe et al. (2013), 'Bringing ecosystem services into economic decision making: Land use in the UK', *Science*, **341**, 45–50.

Conley, D.J. (2012), 'Ecology: Save the Baltic Sea', *Nature*, **486**, 463–464.

De Cian, E., E. Lanzi and R. Roson (2013), 'Seasonal temperature variations and energy demand: A panel cointegration analysis for climate change impact assessment', *Climatic Change*, **116**, 805–825.

Defra (2014), *Government Forestry and Woodlands Policy Statements Incorporating the Government's Response to the Independent Panel on Forestry's Final Report (2013)*, https://www.gov.uk/ government/publications/government-forestry-pol icy-statement (accessed September 2014).

EA-AfA163 (2012), *GQA Headline Indicators of Water Courses (Nutrients)*, Bristol, UK: Environment Agency.

EPA (Environmental Protection Agency) (2012), *Water Quality Standards Handbook*, 2nd edn, EPA-823-B-12-002, Washington, DC: EPA.

European Commission (1991), Directive 91/676/EEC (Nitrates Directive), *Official Journal*, 12 December.

European Commission (2000), Directive 2000/60/EC (Water Framework Directive), *Official Journal*, 22 December.

Fezzi, C. and I.J. Bateman (2011), 'Structural agricultural land use modeling for spatial agro-environmental policy analysis', *American Journal of Agricultural Economics*, **93**, 1168–1188.

Fezzi, C., A.R. Harwood, A.A. Lovett and I.J. Bateman (2015), 'The environmental impact of climate change adaptation on land use and water quality', *Nature Climate Change*, **5**, 255–260.

Goldstein, J.H., G. Caldarone, T.K. Duarte, D. Ennaanay, N. Hannahs, G. Mendoza, S. Polasky, S. Wolny et al. (2012), 'Integrating ecosystem-service tradeoffs into land-use decisions', *Proceedings of the National Academy of Science*, **109**, 7565–7570.

Guo, C. and C. Costello (2013), 'The value of adaptation: Climate change and timberland management', *Journal of Environmental Economics and Management*, **65**, 452–468.

Jarvie, H.P., C. Neal and P.J.A. Withers (2006), 'Sewage-effluent phosphorus: A greater risk to river eutrophication than agricultural phosphorus?', *Science of the Total Environment*, **360**, 246–253.

Johnson, J.A., C.F. Runge, B. Senauer, J. Foley and S. Polasky (2014), 'Global agriculture and carbon trade-offs', *Proceedings of the National Academy of Science*, **111**, 12342–12347.

Kennedy, M. (2009), *Introducing Geographic Information Systems with ArcGIS*, Hoboken, NJ: John Wiley, pp. 433–443.

Kling, C. (2011), 'Economic incentives to improve water quality in agricultural

landscapes: Some new variations on old ideas', *American Journal of Agricultural Economics*, **93**, 297–309.
Lawler, J.J., D.J. Lewis, E. Nelson, A.J. Plantinga, S. Polasky, J.C. Withey, D.P. Helmers, S. Martinuzzi et al. (2014), 'Projected land-use impacts on ecosystem services in the United States', *Proceedings of the National Academy of Science*, **111**, 7492–7497.
Lobell, D.B., W. Schlenker and J. Costa-Roberts (2011), 'Climate trends and global crop production since 1980', *Science*, **333**, 616–620.
McIsaac, G.F., M.B. David, G.Z. Gertner and D.A. Goolsby (2001), 'Eutrophication: Nitrate flux in the Mississippi River', *Nature*, **414**, 166–167.
Nakicenovic, N. and R. Swart (2000), *Special Report on Emissions Scenarios*, Cambridge, UK: Cambridge University Press.
Natural Capital Committee (2014), *The State of Natural Capital: Restoring our Natural Assets*, Second report to the Economic Affairs Committee, London: Defra, http://www.defra.gov.uk/naturalcapitalcommittee/ (accessed September 2014).
Nisson, M., T. Zamparutti, J.E. Petersen, B. Nykvist, P. Rudberg and J. McGuinn (2012), 'Understanding policy coherence: Analytical framework and examples of sector–environment policy interactions in the EU', *Environmental Policy and Governance*, **22**, 395–423.
Patz, J.A., D. Campbell-Lendrum, T. Holloway and J.A. Foley (2005), 'Impact of regional climate change on human health', *Nature*, **438**, 310–317.
Pielke, R., G.P. Prins, S. Rayner and D. Sarewitz (2007), 'Lifting the taboo on adaptation', *Nature*, **445**, 597–598.
Polasky, S., D. Lewis, A. Plantinga and E. Nelson (2014), 'Implementing the optimal provision of ecosystem services', *Proceedings of the National Academy of Sciences*, **111**, 6248–6253.
Sterling, S.M., A. Ducharne and J. Polcher (2013), 'The impact of global land-cover change on the terrestrial water cycle', *Nature Climate Change*, **3**, 385–390.
UKCIP (United Kingdom Climate Impacts Programme) (2011), *UK Climate Projection: Briefing Report*, Exeter, UK: Met Office Hadley Centre.
Vitousek, P.M., H.A. Mooney, J. Lubchenco and J.M. Melillo (1997), 'Human domination of Earth's ecosystems', *Science*, **277**, 494–499.
Willis, K.J. and S.A. Bhagwat (2009), 'Biodiversity and climate change', *Science*, **326**, 806–807.

3. Adaptive capacity contributing to improved agricultural productivity at the household level: empirical findings highlighting the importance of crop insurance

Architesh Panda, Upasna Sharma, K.N. Ninan and Anthony Patt

INTRODUCTION

Few today would argue with the desirability of improving adaptive capacity, especially in developing countries where vulnerability is high and adaptive capacity is low. In that effort, many argue for focusing efforts on the most vulnerable groups, or those with the least adaptive capacity (Hulme et al., 2011). At the national level, there is plentiful data to suggest where these places are. Yohe and Tol (2002), Brooks et al. (2005) and Patt et al. (2010a) have all, for example, conducted cross-sectional analyses to identify the national level indicators that predict large losses of life and property to climate-induced disasters.

But many suggest a need to drill deeper, to target interventions to those households and individuals, within highly vulnerable countries or regions, with the greatest vulnerability or the least adaptive capacity. The World Bank, for example, conducted a study in 2009–10 on the economics of adaptation to climate change (EACC), and in that analysis attempted to identify the most vulnerable groups within individual communities, as a prerequisite to then financing those efforts that would best support those groups. What is interesting from the World Bank study, however, is how inconclusive their findings were. Across several national case studies, there was a great deal of agreement among experts that those with low access to resources were the most vulnerable and least adaptive, and yet little in their quantitative data analysis revealed this to be the case. The authors of one of their case studies, on Mozambique, were even explicit in suggesting that the data did not suggest any significant drivers of differential adaptive capacity existing at the household level. While some households were more likely to engage in a particular adaptation or coping strategy, when they considered multiple adaptation or coping strategies together, they found no significant predictors of difference; it was almost as if, the authors suggested, informal mechanisms to level the playing field for adaptation existed within the communities they were studying (Patt et al., 2010a).

Clearly, a lot of work has gone into identifying the factors that allow people and households to adapt to climate change in particular ways. What the World Bank study highlights, however, is the lack of concrete findings concerning particular interventions that allow people to adapt to climate change in multiple ways. Such ability, we suggest, can be used as an accurate indicator for general adaptive capacity. In this chapter we examine this question, using data from an area of the world that, like Mozambique, is highly vulnerable to multiple climatic hazards and threats. Our case study in Orissa, India, however, allows us to examine the effects of a greater number of potential drivers of differential adaptive capacity, because it is an area where several policy interventions have already taken place, and where the background heterogeneity among households is higher.

BACKGROUND

Adaptation to climate variability among farmers is not a new phenomenon. Farmers have continually responded to weather over the centuries. While farmers have usually adapted to the variability in the climate, climate change poses additional and novel challenges as it is likely to change the baseline of variability itself (Bryant et al., 2000; Adger et al., 2003; Parry et al., 2007). Adaptation is a key factor that will determine the future severity of climate change impacts on agriculture (Brooks et al., 2005; Lobell et al., 2008).

Much of the literature on adaptation and adaptive capacity has considered highly spatially aggregated indicators. Many of these studies have been empirical in nature, attempting to identify the apparent drivers of vulnerability at the national level (Yohe and Tol, 2002; Adger et al., 2004; Brooks et al., 2005; Patt et al., 2010b). Other studies have started with theory, and constructed composite indicators of adaptive capacity or vulnerability at the state, national or regional scale, and then used these indicators to identify the most at-risk places now and in the future. Vincent (2004, 2007), for example, constructed a social vulnerability index, and used this first to assess social vulnerability across African countries, and then to investigate how differences in social vulnerability were highly dependent on the scale of analysis. O'Brien and Leichenko (2000) and O'Brien et al. (2004) constructed an indicator of vulnerability to climate change and economic dependency associated with trade, and used this to map hotspots of vulnerability at the state and district level in India.

A large number of studies consider specific adaptation options themselves, particularly in the agricultural sector (Smit, 1993; Adger et al., 2003; Bradshaw et al., 2004; Orindi and Eriksen, 2005; Bryan et al., 2011). Important options here include crop diversification, mixed crop–livestock farming systems, using different crop varieties, changing planting and harvesting dates, drought-resistant varieties and high-yield water sensitive crops (Bradshaw et al., 2004). Below et al. (2010), who reviewed 17 studies covering data from more than 16 countries in Africa, the Americas, Europe and Asia, found 104 different practices relevant to climate change adaptation. Broadly, these adaptation measures in agriculture can be grouped into four main categories: (1) technological developments, (2) government programmes and insurance, (3) farm production practices, and (4) farm financial management (Smit and Skinner, 2002). Adaptation options in agriculture can also be classified into short-term coping mechanisms and longer-term adaptation measures. Thomas et al. (2007) have classified them into four categories: changes to farming practices as coping by farmers; utilizing the spatial and temporal diversity of the landscape and commercializing of livelihoods as adapting; and using network as both coping and adapting. One of the important limitations of these studies is that they have looked into different types of adaptation options in agriculture and not at the factors affecting the choice of adaptation option at the farm household level.

Several studies have looked at the drivers of differential propensity to engage in one or more of the different options for coping or adapting. Sharma and Patt (2012), for example, studied whether people evacuated their village in response to a warning of a tropical cyclone. They found that on average 65% of the people do so, but also that particular past experiences that individuals had with respect to evacuation influence their choice of whether to evacuate another time. Patt et al. (2005) examined the effects of participatory communication to enhance the use of seasonal climate forecasts in African villages, and found the effect to be a five-fold increase in average adaptation rates. They also found that neither income nor education had a measurable effect on the use of climate forecasts, echoing similar findings by Phillips (2003). In this strand of literature several studies have looked into the factors that influence the choice of particular adaptation options, such as crop selection and livestock management by households (Hassan and Nhemachena, 2008; Kurukulasuriya and Mendelsohn, 2008; Seo and Mendelsohn, 2008; Deressa et al., 2009).

Many of these studies have demonstrated that, just as adaptation options are often location specific, so too are the factors that determine their uptake (Manandhar et al., 2011).

Relatively few studies, however, have considered the generic factors that allow households not to engage in particular options, but rather in a wide range of options. Nhemachena and Hassan (2007) and Hassan and Nhemachena (2008) report on one of these. They used a *logit* regression method to analyse household level data from South Africa, Zambia and Zimbabwe, which included seven farm level adaptation measures and 19 potential drivers of differential adoption. With a sample size of over 1700 respondents, they found almost all of the independent variables to be significant predictors of at least one adaptation option, and many to be significant predictors of multiple options. Of these, several were related to the climate conditions (mean annual temperature and precipitation), while others were community, but not farmer specific, such as living in a property rights regime that granted private rather than communal ownership, or the existence of a local agricultural extensions service office. Among the farmer specific variables, they found that farmer experience, gender of the household head, access to draft power, access to credit and markets, and the pre-existence of mixed cropping strategies all had a significant impact on the likelihood of adopting multiple options. Deressa et al. (2009) reported on a similar analysis based on a sample of over 800 farms in Ethiopia, examining five adaptation options. Like Hassan and Nhemachena, they found the presence of an agricultural extension service to be an important predictor of all five adaptation options, and among farmer specific variables education and farm income predicted four or five options, while access to credit, gender of the household head, and information on climate change each predicted three of the five options.

Patt et al. (2010c) conducted a slightly different analysis in Mozambique, examining both qualitative and quantitative data, although the quantitative element was hampered by a relatively small sample size. The qualitative findings revealed people's opinions about differential adaptive capacity, and suggested a strong belief that low-income and female-headed households have lower adaptive capacity; it is worth noting that the latter opinion contradicts the quantitative findings from Nhemachena and Hassan (2007) and Deressa et al. (2009), both of which showed female-headed households to have higher adaptive capacity. The quantitative findings revealed only one variable, household size, to have a significant effect on one of the adaptation options considered, and none of the variables to predict a propensity to engage in multiple adaptation options. Instead, the researchers found community level indicators, rather than household level ones, to be the important determinants of adaptive capacity measured at the household level.

The studies leave two important questions unanswered. First, can the largely consistent findings of Deressa, Hassan and Nhemachena transfer to a context outside of Africa? While there have been studies of adaptive capacity in the South Asian context, these have been less clearly empirical than the ones in Africa. Aulong et al. (2012) conducted a study in South India, and found that proximity of the administrative centre, information about weather forecasts, the size of the farm, land area rented, education level and farming income seem to distinguish the two groups of farmers with high and low adaptive capacity. Their study, however, did not rest on the purely empirical foundation of the African studies, but rather assessed the effects of the independent variables on a composite index, chosen based on expert-informed theory to be a proxy for adaptive capacity. Jones and Boyd (2011) studied the determinants of adaptive capacity at the household level in Western Nepal, and found a set of social, cognitive and economic determinants. However, their methods were qualitative, and leave open the question of whether quantitative methods would reach the same conclusions; in Patt et al. (2010a) the qualitative and quantitative findings were not in agreement.

Second, the question remains as to whether there are specific interventions that can be taken, other than those that would be undertaken anyway for generic development purposes, which would increase general adaptive capacity at the household level. The studies we have cited show that households with greater access to resources, and greater education, enjoy higher adaptive capacity, and yet achieving higher scores on such indicators is often the goal, not the means, of a great number of development policies and programmes. Improved access to credit is one indicator that could be the subject of an adaptation intervention, although this too is often seen as a method of improving economic development in a country or region, rather than as one tailored only to farming or climate vulnerability. In the study we now report on, we consider a number of potential interventions, many of which are climate specific. We evaluate which of these correspond to consistently increased likelihoods of engaging in multiple adaptation options, and in particular those adaptation options that have the net effect of increasing agricultural productivity.

METHODS

To answer our research questions we developed a structured questionnaire to elicit the adaptation and coping practices used by farm households to cope with drought, as well as a number of factors that could be considered as driving the adaptation actions of the farmers. The study area was in the state of Orissa in Eastern India. The state is an agrarian state with agriculture and animal husbandry contributing to more that 22.6% of the net state domestic product (NSDP) in 2006–07 at 1999–2000 prices and providing employment directly or indirectly to 65% of the total work force (*Status of Agriculture in Odisha*, 2008). The state has a tropical climate, characterized by high temperature, high humidity, medium to high rainfall, and short and mild winters. The normal rainfall of the state is 1451.2 mm. The actual rainfall received varies from district to district. About 84% of rainfall is received during the period from June to September. Even though the quantum of rainfall is quite high, its distribution during the monsoon period is highly uneven and erratic. Floods, drought and cyclones occur often with varying intensities. While the districts in Eastern Orissa suffer from cyclones and floods, the districts of Western Orissa suffer from regular and frequent droughts. The state is divided into ten agro-climatic zones (*State of Environment Report, Orissa*, 2007). Out of the ten agro-climatic zones in Orissa, the survey districts fall under two agro-climatic zones (Western Undulating Zone, Western Central Table Land).

The districts of Western Orissa are counted among the most backward regions in India. Poverty is highest in the southern National Sample Survey (NSS) region followed by the northern region. Poverty in the NSS northern region is as much as 30 percentage points higher than the all-India rural figure of 28.1 at 59.1%. But the southern region has a massive incidence of poverty of 72.7%. Significantly, both of these are the highest for any NSS region in India (FSARO, 2008). The two districts, Bolangir district and Nuapada, where our study areas are located fall under the northern NSS region mentioned above. These two districts are also among the drought prone districts of the country (PACS, 2004). The two districts were chosen purposively to study the impact of droughts and climate variability among the farming households, and both had experienced a severe drought in 2002. These districts have special status under the Revised Long Term Action Plan of the Government of India. They are located in the southwest part of Orissa between latitudes 17°48'30"N and 21°8'45"N and longitudes 81°22'48"E and 84°15'32"E in the Eastern Ghats range. The region belongs to the sub-humid temperate region of India with an average rainfall ranging from 1100 to 1400 mm.

Bolangir district is located in the Western Central Table Land zone of Orissa. It covers an area of some 6924.3 square kilometres with a population of 1 336 000, and a population density of 193 per square kilometre (Census of India, 2001). The district receives a normal annual average rainfall of 1443.5 mm. Nuapada adjoins Bolangir district to the west and is itself the furthest central west district in Orissa, bordered by Madhya Pradesh, and lies between 20°0'N and 21°5'N latitudes and between 82°20'E and 82°40'E longitudes. It covers an area of some 3407.5 square kilometres with a population of 530000 and a population density of 188 per square kilometre (Census of India, 2001). Agro-climatic data indicate three rainfall zones, with lower rainfall to the south of the district (below 1000 mm), a higher belt in the middle (above 1200 mm) and 1000–1200 mm occurring to the northern quarter and a small section, some 10% of land area, to the south-eastern edge. Two villages from each district – Rengali and Budibahal in the Khaprakhol block in Bolangir district and Kirkita and Khamtari villages in Kharaiar block in Naupada district – were chosen as our study areas.

The dependent variables, representing adaptations taken, were based on previous qualitative interactions with stakeholders in the study areas. We based the individual questions concerning independent variables on a review of existing theory (Bradshaw et al., 2004; Maddison, 2006; Thomas et al., 2007; Kurukulasuriya and Mendelsohn, 2008; Deressa et al., 2009; Below et al., 2010, Bryan et al., 2011; Manandhar et al., 2011). Table 3.1 lists the variables for which we collected data.

We conducted a household survey in the four villages during January to May in 2010. The unit of analysis is household and stratified random sampling method was used to choose the households on the basis of large, small and marginal farmers in the villages. Initially 200 households, i.e. 50 households from each village, were surveyed but due to non-response of certain parts in the questionnaire we had to drop 17 questionnaires, thus making our total sample size 183. We then applied the relevant statistical technique (*logit* regression) to study the relationship between adaptation actions and their drivers.

RESULTS

The farmers in our sample reported using each of the adaptation options that we had queried them on. Figure 3.1 shows the frequency with which the different options were employed, revealing that water conservation practices were the least frequently employed, and changing planting dates the most frequently employed. Half of these practices – changing planting dates, shifting from rice to cotton, reducing rain dependence through the use of wells or water storage, and water conservation – all have the effect of increasing yields, holding climate constant. The other half of the practices – use of early maturing varieties, diversification of varieties, reducing cultivated land, and income diversification – all have the effect of

Table 3.1 Description of independent variables queried in the survey instrument

Independent variable	Description
Monetary drought loss	Continuous
Total land holding	Continuous
Total income	Continuous
Agricultural income	Continuous
Non-farm income	Binary (yes, no)
Below poverty line (BPL) category	Binary (yes, no)
Size of the household	Numerical
Level of education	Categorical
Farming experience (years)	Numerical
Number of dependent persons	Numerical
Number of persons migrated	Numerical
Area under low land	Continuous
Area of high land	Continuous
Area irrigated	Continuous
Possession of well	Binary (yes, no)
Access to additional water resources	Binary (yes, no)
Access to climate information	Binary (yes, no)
Membership in self-help group	Binary (yes, no)
Access to credit	Binary (yes, no)
Access to crop insurance	Binary (yes, no)
Perception on changing climate	Categorical
Perception of frequency of drought	Categorical
Perception on declining rainfall	Categorical
Perception on the first rain	Categorical

lowering the risks associated with a poor harvest, but do not have the effect of increasing yields, and indeed on average lead to lower yields.

Next, we analysed the relationships between the independent variables and each of the adaptation strategies shown in Figure 3.1. We conducted a separate *logit* regression for each dependent variable (i.e. the adaptation strategies). In each case, given a sample size of 183, we preserved degrees of freedom by dropping those independent variables that on first inspection gave insignificant results, and for which there was no prior expectation, based on theory, of a causal relationship. We used the *logit* regression results to calculate odds ratios. An odds ratio of 1 suggests that the independent variable has no effect on the dependent variable. An odds ratio below 1 suggests a negative correlation; for example, a value of 0.5 would indicate that increasing the independent variable by one unit would result in observing the adaptation option only 50% as frequently. An odds ratio above one

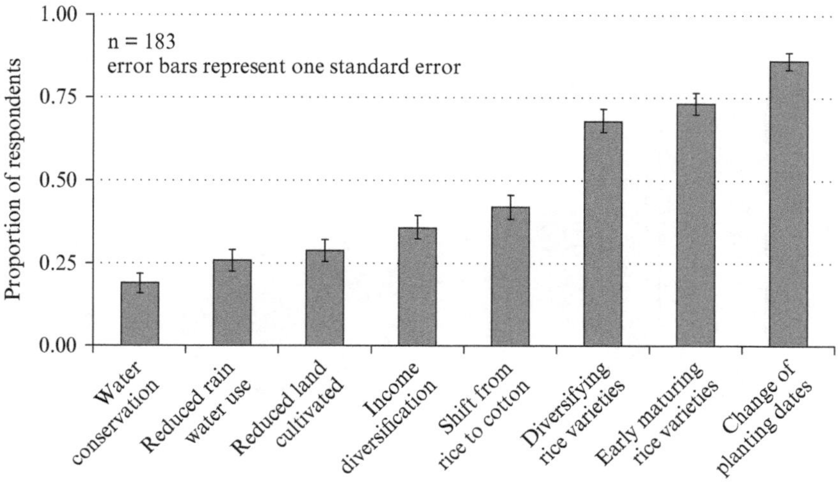

Figure 3.1 Frequency of adaptation options queried

suggests a positive correlation; for example, a value of 2 would indicate that increasing the independent variable by one unit would make it twice as likely to observe the adaptation option. The results appear in Table 3.2.

As seen in Table 3.2, we identified significant relationships between many of the independent variables and one or more of the adaptation actions. Having experienced previous drought loss, land area under cultivation, family size, migration status, being in the high land, additional water access and perceived change in the time of the first rainfall all correlated significantly with a single adaptation strategy. Of these, additional water access had the strongest single positive effect on an adaptation option, with an odds ratio of over 8. Income variables – total income, agricultural income, and being above the poverty threshold – all correlated with the same two adaptation options, namely shifting from rice to cotton, and taking advantage of early maturing rice varieties. Farming experience, the area irrigated, access to credit, and the perception that the frequency of drought had changed correlated with two adaptation options. Having access to crop insurance, and having a well as a source of water, correlated with three adaptation options each. Interestingly, several of the indicators that were expected to be significant were not. These included education, non-farm income, the number of dependants, access to climate information, membership in a self-help group, perception that the climate is changing, and perception that there had been an overall decline in the amount of rainfall.

Our final analysis was to identify those indicators that had a consistently

Table 3.2 Odds ratios derived from logit regression results

	Water conservation	Reduced rain use	Shift from rice to cotton	Change of planting dates	Reduced cultivated land	Income diversification	Diversifying rice varieties	Early maturing rice varieties
Drought loss	1.01	0.98	1.02	1.04	1.01	1.00	0.98	1.09**
Land	1.06			1.46**	1.08	1.02	0.97	
Total income	1.02	0.97	1.27**	0.91	1	0.94	1.07	1.63**
Agricultural income	1.00	1.17	0.15***				0.97	0.56**
Non-farm income					0.58			
Below poverty line	1.08	0.71	0.15***				0.88	0.25***
Family size	0.80*	0.97	0.64					
Education	1.19	0.77	1.54	1.31	0.99	0.95	0.97	0.70
Farming experience	1.00		0.98	0.95***	1.01		1.04**	0.99
Number of dependants			1.07			1.07		
Migration					1.00	1.39**		
Bahal (low land)	1.59**	1.06	0.90	0.58*	1.03		1.26	
Aat (high land)	0.88	0.88	1.64***		0.93		1.12	
Area irrigated	0.57**	1.99	2.35	0.58	0.71		0.57**	0.93
Well	3.51**	9.86***	1.14				1.45	2.76**
Additional water access					1.03		8.38**	

Climate information	0.72	1.86	0.78			0.92	0.91	
Self-help group	1.11	0.73				1.64	0.50	
Access to credit	0.93	0.30**	0.37*		1.37	0.74	1.05	1.33
Crop insurance	0.61	2.50**	6.55***		0.41**	0.46**	0.56	
Perceived change in general climate	2.14		0.7					
Perceived drought frequency changes	0.37**		1.03	2.18	1.06	2.68**	1.04	1.04
Perceived decline in rainfall	0.98		0.56		1.12	0.83	0.41	0.66
Perceived change in time of first rain				1.06***		5.27		0.84

Notes:
Empty boxes represent variables that were dropped in order to preserve degrees of freedom.
Asterisks indicate statistically significant results at *** 1%, ** 5% and * 10% significance level.
N = 183.

positive impact on adaptation, in such a way as to improve yields. For policy makers, we expect these indicators to be especially important, as they could represent key interventions. Figure 3.2 shows a 'heat diagram', in which we do this. It shows those indicators that we identified having a significant effect on at least two adaptation options. We then colour code the direction and magnitude of the effect, distinguishing between those adaptation options that have a net yield-increasing effect, and those that have a net yield-decreasing effect. For the former, black/white horizontal lines indicate a strong positive effect, light grey a weak positive effect, black/white vertical lines a weak negative effect, and dark black a strong negative effect. For the latter, we reverse the colours. Hence, black/white horizontal lines indicate either a strongly positive effect on an adaptation option that increases yields, or a strongly negative effect on an adaptation option that decreases yields. What we observe is that a single indicator – access to crop insurance – has a consistent effect on two or more adaptation options in a manner that tends to increase yields. Access to a well has a yield-increasing effect in two places, but a yield-decreasing effect in another. Six indicators have roughly balanced effects. Three indicators – farming experience, access to credit and the perception of change in drought frequency – have effects that would be consistently negative on yields.

DISCUSSION

If we define adaptive capacity in the agricultural sector as the propensity to adopt farming practices that maintain higher yields in the face of climate change, what differentiates households with high adaptive capacity from those with low adaptive capacity? Our results indicate that it is the ability to take risks. In particular, we see this in two ways. First, households with a good source of water for irrigation can take the risk of planting water intensive crops, since if needed they can substitute with hard work by carrying or pumping water for lack of adequate rainfall. Second, households with access to insurance can plant more and higher-yielding crops, since if the crops fail they will have recourse to insurance pay-outs. Of these two, the effects of crop insurance appear to be stronger. As such, they provide important support to pilot programmes to develop crop insurance in ways that do in fact represent a good financial deal for small-holder farmers (Freeman and Kunreuther, 2002; Leftley and Mapfumo, 2006; Barnett and Mahul, 2007; Chantarat et al., 2007; Patt et al., 2010c), and place a premium on ways to make such insurance financially sustainable and more attractive to farmers (Carter et al., 2008; Hochrainer et al., 2009; Meze-Hausken et al., 2009; Patt et al., 2009).

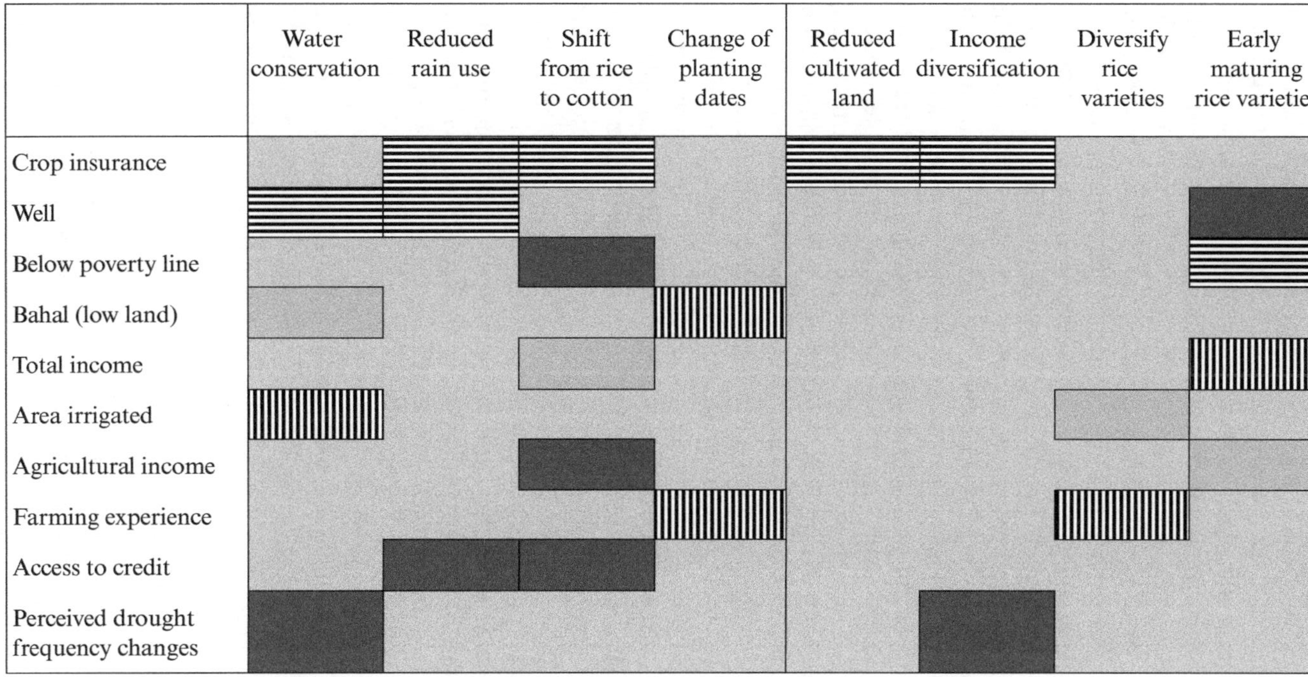

Note: The first four adaptation options increase yields, and, for these, black/white horizontal lines are used to represent odds ratios greater than 2, light grey between 1 and 2, black/white vertical lines between 0.5 and 1, and dark black below 0.5. The second four adaptation options have the effect of reducing yields, and for these the colour coding is reversed.

Figure 3.2 'Heat diagram' showing significant odds ratios

Our results are also interesting for the indicators that they suggest do not consistently improve adaptive capacity. Surprisingly we see no significant effects for education or the presence of climate information, although this is consistent with studies such as Patt et al. (2005), where education was found not to be significant, and climate information rarely adopted in the absence of participatory communication practices of a type not present in our study area. We observed farming experience to have an effect, but one that tended to decrease yields; more experienced farmers, perhaps, know how to cut their losses. Access to credit has been observed to be an important factor for adaptation in Africa (Nhemachena and Hassan, 2007; Deressa et al., 2009), and yet we did not observe the effects to be good. In our case, it enabled rice farmers to continue to farm rice, rather than switch to cotton, and to avoid having to find ways to lessen their dependency on rain water. The perception that drought is increasing in frequency, was surprisingly associated with a marked decrease in the use of water conservation practices, although this could be a case of the causal arrow going in the other direction: farmers who did not engage in water conservation would deny seeing an increased need to do so. Several other variables we saw having mixed effects, and not inconsistent with the past literature.

Our results need to be interpreted with caution, in two respects. The first has to do with the limitation of empirical design. In comparison with other studies showing strong determinants of adaptive capacity, our sample size was smaller; that could lead us to fail to identify important determinants that are in fact there. The second has to do with the distinction we make between adaptation options that increase or decrease yields, respectively. We anticipate that policy makers care about improving adaptation in ways that increase yields, since this could, if coupled with adequate distribution, improve the food security situation, or at least self-sufficiency, of the region under consideration. On the other hand, adaptation options that decrease yields while increasing something else, such as income diversification, could be the best possible options for many individual farmers and, to the extent that a region is well connected to food markets, could improve the food security of the region simply by improving the financial means to access those markets. Measures that purely reduce risk – diversifying crop varieties and moving towards shorter season varieties – will not improve average conditions, but could have an important effect of reducing the need for emergency intervention in the case of drought.

Finally, our results suggest that continued caution is warranted in terms of the assumptions that we make about differential adaptive capacity. We started the chapter with a cautionary tale from a World Bank study, in which the quantitative evidence appeared to provide less support for particular factors enhancing or detracting from adaptive capacity than popular opinion held to be the case (Patt et al., 2010a). Our conclusions are somewhat consistent with this. When one looks hard at data within particular communities, it is quite difficult to conclude that differences between households have a marked effect on those households' capacity to engage in adaptation. Instead, it may well be that particular communities, because of their location and resource endowments, or their access to information and finance, may be in a better position to adapt than others. More research is certainly necessary to throw light on this.

ACKNOWLEDGEMENTS

Funding for the field work component of this research, as a part of the PhD work, came from Institute for Social and Economic Change (ISEC), India. Funding for the analysis of the results and the drafting of this chapter came from the Young Scientists Summer Program of the International Institute for Applied Systems Analysis, Vienna with the lead author's participation supported by the Technology Information, Forecasting and Assessment Council (TIFAC) of India. Any errors of fact or interpretation are those of the authors.

REFERENCES

Adger, W.N., S. Huq, K. Brown, D. Conway and M. Hulme (2003), 'Adaptation to climate change in the developing world', *Progress in Development Studies*, **3**, 179–195.

Adger, W.N., N. Brooks, G. Bentham, M. Agnew and S. Eriksen (2004), *New Indicators of Vulnerability and Adaptive Capacity*, Norwich, UK: Tyndall Centre for Climate Change Research.

Aulong, S., B. Chaudhuri, L. Farnier, S. Galab, J. Guerrin, H. Himanshu and P. Prudhvikar Reddy (2012), 'Are South Indian farmers adaptable to global change? A case in an Andhra Pradesh catchment basin', *Regional Environmental Change*, **12** (3), 423–436.

Barnett, B.J. and O. Mahul (2007), 'Weather index insurance for agriculture and rural areas in lower-income countries', *American Journal of Agricultural Economics*, **89** (5), 1241–1247.

Below, T., A. Artner, R. Siebert and S. Sieber (2010), *Micro-Level Practices to Adapt to Climate Change for African Small-Scale Farmers: A Review of Selected Literature*, IFPRI Discussion Paper 00953, February, Washington, DC: International Food Policy Research Institute.

Bradshaw, B., H. Dolan and B. Smit (2004), 'Farm-level adaptation to climatic variability and change: Crop diversification in the Canadian prairies', *Climatic Change*, **67**, 119–141.

Brooks, N., W.N. Adger and P.M. Kelly (2005), 'The determinants of vulnerability

and adaptive capacity at the national level and the implications for adaptation', *Global Environmental Change*, **15** (2), 151–163.
Bryan, E., C. Ringler, B. Okoba, C. Roncoli, S. Silvestri and M. Herrero (2011), 'Adapting agriculture to climate change in Kenya: Household and community strategies and determinants', Paper presented at the international conference on the Economics of Adaptation to Climate Change in Low-Income Countries, 18–19 May, http://www.gwu.edu/~iiep/adaptation/docs/Bryan,%20Adapting%20Agriculture%20to%20Climate%20Change%20in%20Kenya%20Household%20and%20Community.pdf.
Bryant, C.R., B. Smit, M. Brklacich, T.R. Johnston, J. Smithers, Q. Chjotti and B. Singh (2000), 'Adaptation in Canadian agriculture to climate variability and change', *Climatic Change*, **45**, 181–201.
Carter, M., C.B. Barrett, S. Boucher, S. Chantarat, F. Galarza, J.G. McPeak, A. Mude and C. Trivelli (2008), 'Insuring the never before insured: Explaining index insurance through financial education games', University of Wisconsin.
Census of India (2001), *State Primary Census Abstract*, New Delhi: Government of India, Ministry of Home Affairs.
Chantarat, S., C.B. Barrett, A. Mude and C. Turvey (2007), 'Using weather index insurance to improve drought response for famine prevention', *American Journal of Agricultural Economics*, **89** (5), 1262–1268.
Deressa, T.T., R.M. Hassan, C. Ringler, T. Alemu and M. Yesuf (2009), 'Determinants of farmers' choice of adaptation methods to climate change in the Nile Basin of Ethiopia', *Global Environmental Change*, **19** (2) (May), 248–255.
FSARO (*Food Security Atlas of Rural Orissa*) (2008), New Delhi: Institute for Human Development.
Freeman, P.K. and H. Kunreuther (2002), 'Environmental risk management for developing countries', *Geneva Papers on Risk and Insurance*, **27** (2), 196–214.
Hassan, R. and R. Nhemachena (2008), 'Determinants of African farmers' strategies for adapting to climate change: Multinomial choice analysis', *African Journal of Agricultural and Resource Economics*, **2** (1) (March), 83–104.
Hochrainer, S., R. Mechler and G. Pflug (2009), 'Climate change and financial adaptation in Africa: Investigating the impact of climate change on the robustness of micro-insurance in Malawi', *Mitigation and Adaptation Strategies for Global Change*, **14** (3), 231–250.
Hulme, M., S.J. O'Neill and S. Dessai (2011), 'Is weather event attribution necessary for adaptation funding?', *Science*, **334** (6057) (11 November), 764–765.
Jones, L. and E. Boyd (2011), 'Exploring social barriers to adaptation: Insights from Western Nepal', *Global Environmental Change*, **21** (4) (October), 1262–1274.
Kurukulasuriya, P. and R. Mendelsohn (2008), 'A Ricardian analysis of the impact of climate change on African cropland', *African Journal of Agricultural and Resource Economics*, **2** (1), 1–23.
Leftley, R. and S. Mapfumo (2006), *Effective Microinsurance Programs to Reduce Vulnerability*, Opportunity International Network.
Lobell, B.D., M.B. Burke, C. Tebaldi, M.D. Mastrandrea, W.P. Falcon and R.L. Naylor (2008), 'Prioritizing climate change adaptation needs for food security in 2030', *Science*, **319** (1) (February), 607–610.
Maddison, D. (2006), *The Perception of and Adaptation to Climate Change in Africa*, CEEPA Discussion Paper No. 10, Pretoria: Centre for Environmental Economics and Policy in Africa.
Manandhar, S., D. Schmidt Vogt, S.R. Perret and F. Kazama (2011), 'Adapting

cropping systems to climate change in Nepal: A cross-regional study of farmers' perception and practices', *Regional Environmental Change*, **11**, 335–348.

Meze-Hausken, E., A. Patt and S. Fritz (2009), 'Reducing climate risk for micro-insurance providers in Africa: A case study of Ethiopia', *Global Environmental Change*, **19**, 66–73.

Nhemachena, C. and R.M. Hassan (2007), *Micro-Level Analysis of Farmers' Adaptation to Climate Change in Southern Africa*, IFPRI Discussion Paper 00714, August, Washington, DC: International Food Policy Research Institute and Centre for Environmental Economics and Policy in Africa.

O'Brien, K. and R. Leichenko (2000), 'Double exposure: Assessing the impacts of climate change within the context of economic globalization', *Global Environmental Change*, **10**, 221–232.

O'Brien, K., R. Leichenko, U. Kelkar, H. Venema, G. Aandahl, H. Tompkins, A. Javed, S. Bhadwal et al. (2004), 'Mapping vulnerability to multiple stressors: Climate change and globalization in India', *Global Environmental Change*, **14**, 303–313.

Orindi, V.A. and S. Eriksen (2005), *Mainstreaming Adaptation to Climate Change in the Development Process in Uganda*, Ecopolicy Series 15, Nairobi: African Centre for Technology Studies (ACTS).

PACS (2004), *Drought in India: Challenges and Initiatives*, New Delhi: Poorest Areas Civil Society (PACS) Programs, http://www.empowerpoor.org/downloads/drought1.pdf (accessed 23 November 2011).

Parry, M.L., O.F. Canziani, J. Palutikof, P. van der Linden and C. Hanson (2007), *Climate Change 2007: Impacts, Adaptation and Vulnerability: Contribution of Working Group II to the Fourth Assessment Report of the Intergovernmental Panel on Climate Change*, Cambridge, UK: Cambridge University Press.

Patt, A.G., Pablo Suarez and Chiedza Gwata (2005), 'Effects of seasonal climate forecasts and participatory workshops among subsistence farmers in Zimbabwe', *Proceedings of the National Academy of Sciences of the United States of America*, **102**, 12623–12628.

Patt, A.G., N. Peterson, M. Carter, M. Velez, U. Hess and P. Suarez (2009), 'Making index insurance attractive to farmers', *Mitigation and Adaptation Strategies for Global Change*, **14**, 737–753.

Patt, A.G., R. Varela, I. Nhantumbo and L. Bizikova (2010a), *The Social Dimensions of Adaptation to Climate Change in Mozambique*, Washington, DC: World Bank.

Patt, A.G., M. Tadross, P. Nussbaumer, K. Asante, M. Metzger, J. Rafael, A. Gujon and G. Brundrit (2010b), 'Estimating least-developed countries' vulnerability to climate-related extreme events over the next 50 years', *Proceedings of the National Academy of Sciences*, **107**, 1333–1337.

Patt, A.G., P. Suarez and U. Hess (2010c), 'How much do smallholder farmers understand and want insurance?', *Global Environmental Change*, **20**, 153–161.

Phillips, J. (2003), 'Determinants of forecast use among communal farmers in Zimbabwe', in Karen O'Brien and Coleen Vogel (eds), *Coping with Climate Variability: The Use of Seasonal Climate Forecasts in Southern Africa*, Aldershot, UK: Ashgate, pp. 110–128.

Seo, S.N. and R. Mendelsohn (2008), 'Measuring impacts and adaptations to climate change: A structural Ricardian model of African livestock management', *Agricultural Economics*, **38**, 1–15.

Sharma, U. and A. Patt (2012), 'Disaster warning response: The effects of different types of personal experience', *Natural Hazards*, **60** (2) (1 January), 409–423.

Smit, B. (ed.) (1993), *Adaptation to Climatic Variability and Change*, Guelph, ON: Environment Canada.

Smit, B. and W.M. Skinner (2002), 'Adaptation options in agriculture to climate change: A typology', *Mitigation and Adaptation Strategies for Global Change*, **7**, 85–114.

State of Environment Report, Orissa (2007), Bhubaneswar, India: Government of Odisha.

Status of Agriculture in Odisha (2008), Bhubaneswar, India: Department of Agriculture and Farmers Welfare, Government of Odisha.

Thomas, D.S.G., C. Twyman, H. Osbahr and B. Hewitson (2007), 'Adaptation to climate change and variability: Farmer responses to intra-seasonal precipitation trends in South Africa', *Climatic Change*, **83**, 301–322.

Vincent, K. (2004), *Creating an Index of Social Vulnerability to Climate Change for Africa*, Norwich, UK: Tyndall Centre for Climate Research.

Vincent, K. (2007), 'Uncertainty in adaptive capacity and the importance of scale', *Global Environmental Change*, **17** (1), 12–24.

Yohe, G. and R.S.J. Tol (2002), 'Indicators for social and economic coping capacity – moving towards a working definition of adaptive capacity', *Global Environmental Change*, **12**, 25–40.

4. Adapting to climate change and improving urban resilience: the role of nature and biodiversity protection in cities

Konar Mutafoglu, Patrick ten Brink, Sabrina Dekker, Jamie Woollard and Jean-Pierre Schweitzer

INTRODUCTION – HEAT STRESS AS A PUBLIC HEALTH CHALLENGE

Urban populations all over the world are exposed to the urban heat island (UHI) effect and can experience heat stress, potentially exacerbated by rising average and extreme temperatures in the course of climate change. This can pose a serious public health challenge, as recent experience with heatwaves shows. Nature can play an important role in improving micro-climatic conditions in cities and help improve resilience against undesirable public health impacts. This chapter discusses the role of nature in urban and suburban settings and its role in building resilience against heat stress.

The remainder of this section will present the challenges that cities face. Urban heat islands are widespread in densely populated and built-up areas. Moreover, changing climatic conditions give rise to the concern that heat stress will gain in importance to public health. The second section shows how nature-based solutions can improve micro-climatic conditions in cities, including a number of applications and examples. The third section looks more closely at several governance aspects to understand how cities and regions can make better use of nature to address heat stress and build resilience against extreme climatic conditions such as prolonged heatwaves. The last section draws a number of conclusions and makes suggestions for a way forward.

Urban Heat Islands and Heat Stress

The ability of buildings to retain heat, combined with a high concentration of buildings in urban areas, results in urban heat islands (Watkins et al., 2007). Urban heat islands can be observed in cities all over the world, while the heat differential between the built-up urban area and suburban areas differs with the degree of green coverage in urban centres (Zoulia et al., 2009; Peng et al., 2012).

Heat stress occurs when the human body is unable to regulate its normal temperature. If left untreated, heat stress progresses to heat exhaustion, then to heat stroke, to hyperthermia and, ultimately, mortality (Kovats et al., 1999; Kovats and Ebi, 2006). Children, elderly people and those who work in hot environments are especially susceptible. Individuals with pre-existing health conditions, such as cardiovascular and respiratory diseases, are also at greater risk (Baccini et al., 2008; Hallegatte et al., 2011). In Europe, UHIs can increase urban temperatures by up to 12 degrees compared to non-urban areas (Depietri et al., 2011). The 2003 heatwave caused up to 70 000 deaths over four months across Europe, many of which could have been prevented (EEA, 2012).

Additional Impact of Climate Change

Climate projections suggest that the prevalence of heatwaves will increase in the future; thus the risks of heat stress will rise (IPCC, 2014). For most parts of Europe, for example, climate projections conclude that average and extreme temperatures are likely to increase (Ciscar et al., 2014; IPCC, 2014). Around 75 per cent of the European population live in urban areas, which will be exposed to rising temperatures from climate change (EEA, 2012). The EEA, JRC and WHO (2008) projected almost 86 000 additional deaths per year in 2071–2100 in the EU-27 member states compared to the 1961–1990 EU-25 average, based on a severe scenario for temperature increase.

In addition, heat stress can affect economic output, as it affects people's productivity, especially those who work outdoors. Hübler et al. (2008) estimate heat-induced output losses in Germany of between 0.1 and 0.5 per cent of GDP, or €2.5 to €10.4 billion per annum by the end of the 21st century. The risks of heat stress are also unequally distributed according to geographic and social factors. Age, gender and income are often determinants of vulnerability. For example, during the 2003 heatwave in France, mortality rates from heat stress were twice as high in the most deprived provinces (Rey et al., 2009).

The Role of Nature for Mitigating Heat Stress

Scientific evidence confirms that nature can contribute to improving urban climatic conditions overall and can provide areas of cooling, that is, 'cool oases', by regulating local climate (WHO, 2005). This is one of the services that ecosystems can offer, whether urban or suburban protected areas or urban green infrastructure such as tree-lined streets, parks or green roofs (see Figure 4.1). A growing range of institutions and stakeholders are starting to appreciate the multiple benefits of nature. Policies and instruments to help harness the benefits are being put in place to build synergies between nature and other objectives and to avoid unacceptable trade-offs, such as loss of green space in cities where these provide important services to citizens, many of which were not previously accounted for (Tzoulas et al., 2007; ten Brink et al., 2016).

While this chapter focuses primarily on the benefits of nature and green infrastructure in addressing extreme temperatures, nature provides a much wider array of ecosystem services (TEEB, 2011). Green infrastructure can, for this reason, service multiple objectives and deliver multiple benefits. For instance, a policy of nature protection to preserve green infrastructure for urban cooling will most likely provide ancillary benefits, such as reduction of exposure to air pollution, biodiversity conservation, or even employment and urban regeneration (ten Brink et al., 2016). Likewise, greening without the aim of micro-climatic regulation may unintentionally provide these services.

THE ROLE OF NATURE IN REDUCING THE HEAT ISLAND EFFECT IN CITIES

There exists a long tradition of using green spaces to provide cooling benefits in and around cities. On a summer day, an urban park with a dense canopy will typically be several degrees cooler than its surrounding built environment. Green recreational areas allow for weekend escapes from densely developed cities. Large, interconnected, undeveloped areas of land generate cool air that can help moderate temperatures in city centres and mitigate heat stress. With expected rising average and extreme temperatures associated with climate change, the cooling and thermal comfort benefits of nature are gaining importance (Jim, 2013). This section looks at the mechanisms at work and presents a number of case examples from across Europe.

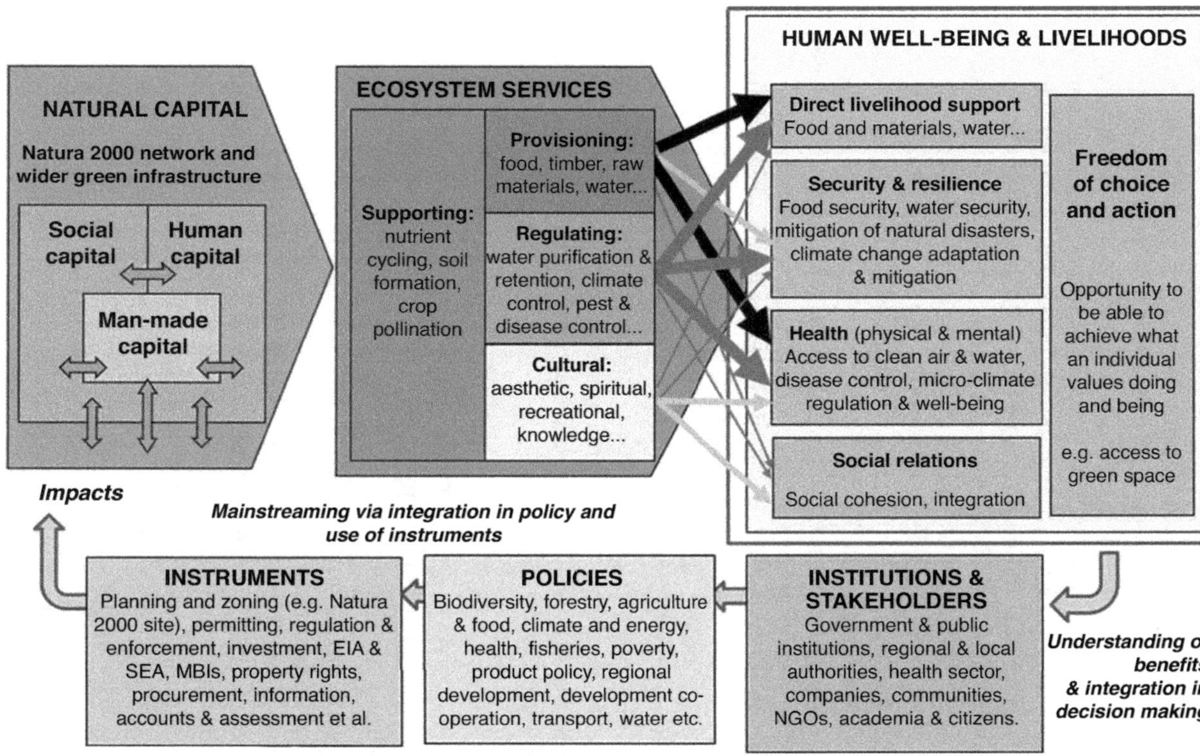

Source: Own representation building on MA (2005) and TEEB (2011).

Figure 4.1 Human well-being benefits from ecosystem services and institutional responses

How Does Nature Help in Cooling Cities?

Protected areas, green open spaces like parks and fields, and various forms of green urban infrastructure (green roofs, green walls and green belts) can help mitigate the UHI effect, the expected rise in average temperatures, and the impacts of heatwaves (Susca et al., 2011). Two processes are at work. First, green infrastructure provides shade that reduces the absorption and retention of heat by surfaces in the built environment. Second, plants and trees cool their environment through evapotranspiration (Ennos, 2012). The magnitude of cooling is dependent on the configuration, type, size, health and density of vegetation (Zupancic et al., 2015). Seasonal and temporal variations may also influence the cooling capacity of vegetation (Renaud and Rebetez, 2009).

This effect can be observed close to the green area itself but also in the vicinity. Ideally, such measures should be combined with activities to manage storm water and increase the permeability of soils, which further support cooling through evaporation. Blue infrastructure, namely bodies of water such as rivers, is also shown to contribute to reducing the heat island effect, including air temperatures at night (Hathway and Sharples, 2012).

Traditional Green Infrastructure Elements – Parks, Trees, Green Spaces

Historically, green infrastructure has focused on parks of various sizes and street trees, and these continue to be important features of the urban environment providing public health benefits. Large parks and protected areas make significant contributions to cooling and provide oases on hot days (Bowler et al., 2010). The cooling effect of parks may extend to the wider surrounding area. For example, a long-running study of three parks in Gothenburg, Sweden showed that the effect could reach as far as 1 km from the park boundary, for the largest park that was observed (156 ha) (Upmanis et al., 1998). In US cities such as Los Angeles, municipal health and emergency services keep parks open for a longer time during heatwaves. They are advertised to the public as 'cooling centres' alongside air-conditioned public spaces, recognising that they may be cooler than people's homes.

Small urban green patches can also make significant contributions to local cooling. For example, a study of a 0.24 ha urban park in Lisbon recorded temperatures of up to 6.9 °C cooler than surrounding areas on the hottest days (Oliveira et al., 2011). In a study of the National Garden in Athens, Greece, Zoulia et al. (2009) investigated the temperature differences along various routes within the park at different times throughout the day. Their findings suggest that the density of vegetation and type play a role in decreasing temperatures in the surrounding area.

A study in Manchester, UK showed that an increase of green areas by 10 per cent would keep the maximum temperatures by 2080 at nearly the same level as 1961–1990 baseline conditions and mitigate an expected temperature rise of 4 °C (Gill et al., 2007).

Several studies have examined the benefits of trees in contrast to alternative forms of green infrastructure (e.g. green roofs or green walls), largely finding trees to be the most effective in providing shade and cooling (Bowler et al., 2010; Teuling et al., 2010). Furthermore, larger green spaces demonstrated greater cooling benefits.

Green Roofs and Green Walls

Green roofs and green walls provide valuable additions to urban green infrastructure and can help to improve climatic conditions in cities. The advantages of these approaches are that they do not compete in the same way with other land uses in urban spaces. In town centres and high-density residential areas, green roofs have been shown to provide an effective means for reducing surface temperatures and enhancing the thermal efficiency of buildings (Handley and Carter, 2006; Carter and Butler, 2008).

In Manchester, Red Rose Forest support the implementation both of 'big green roofs', including a 750 m^2 roof at Manchester Metropolitan University (MMU, 2011), and also of 'small green roofs', for instance on garden sheds (Red Rose Forest, 2006). Even small roofs can act as hotspots for biodiversity, provide insulation and absorb radiation.

Combined Green and Blue Infrastructure

Whilst green infrastructure is often associated only with vegetation, it also includes 'blue' water infrastructures. Indeed vegetation and water are linked inexorably, as are the climate benefits related to green infrastructure, such as the cooling function from evapotranspiration.

Increasingly, green infrastructure strategies are being explored that integrate both green and blue infrastructure in relation to climate change. For example, the Blue Green Dream research project is working to enhance the synergy of urban blue and green systems and provide effective, multifunctional blue–green solutions to support urban adaptation to future climatic changes. This has included ten demonstration sites in the UK, the Netherlands, France and Germany (Blue Green Dream, 2015). Lyon in France has gone further and invested in green–blue infrastructure (see Box 4.1).

BOX 4.1: LYON, FRANCE – THE RHÔNE RIVER, RAIN WATER AND COPING WITH HEATWAVES

Lyon has had to tackle multiple challenges in relation to climate change. The 2003 heatwave increased mortality in the city by 80 per cent, above the national average for a French city (Vandentorren et al., 2004). Green and blue infrastructure, including the Rhône River, has been at the heart of the solution. Increasingly, the city is making the links between the benefits of green and blue infrastructure, conservation and practical urban spaces.

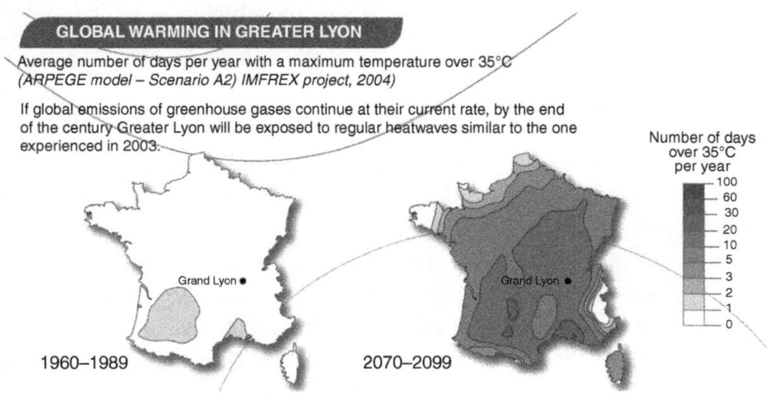

Source: http://www.ale-lyon.org/download/Plan_climat_Grand_Lyon/plaquette_PCT_GL_UK.pdf.

In the 20th century, a focus on grey infrastructural solutions to flooding in the city left both a disconnect between the citizens and their river and a number of issues in relation to future risks for the city. Several green infrastructure projects in the city have introduced a shift from grey to green infrastructure in city planning. The Greater Lyon initiative includes a shade indicator, calculating annually the area of shade provided by trees in public spaces. Such a quantitative indicator allows clear targets to be developed in the future (Grand Lyon, 2007a).

The city's development is increasingly guided by strategies such as the Water Strategy (Grand Lyon, 2007a) and the Climate Plan (Grand Lyon, 2007b). The strategies stress that, whilst flooding has been a priority of the past, increasingly water scarcity and summer heatwaves will affect the city in the future. In 2007, the city reopened public access to the banks of the city's river (Grand Lyon, 2014). The €42 million redevelopment programme, Berges du Rhône, replaces a bitumen car park with 5 km of pathway and green spaces, as well as providing soft flood protection and opportunities for recreation, and connecting two urban parks and the protected Natura 2000 site Grand Parc de Miribel-Jonage in the north of the city.

REALISING PUBLIC HEALTH BENEFITS THROUGH INVESTING IN NATURE: GOVERNANCE, STAKEHOLDER COLLABORATION AND SUPPORTING INSTRUMENTS

Addressing the impacts of climate change, in particular rising temperatures and urban heat islands, requires action by a broad range of stakeholders at local, regional, national and supra-national levels. This section presents and discusses some of the supporting instruments and governance mechanisms that are currently in place to address the impacts of climate change on human health.

Regulation and Legislation

The case examples presented thus far demonstrate that mitigating urban heat island effects typically results from local action. However, local action often requires the support of regional and national governments and supra-national institutions (e.g. the European Union) and international governing bodies. Through the introduction of regulation and legislation, they create an environment that makes way for local governments to create plans and policies that promote green infrastructure and protect nature as a priority. For European cities, the European Union (EU) empowers local governments through directives and regulations that commit member states to achieving targets, as well as through strategies and objectives to guide countries. Moreover, the EU supports a number of initiatives that enable local governments to interact with each other and share knowledge and experience, such as the Covenant of Mayors initiative. It is in this context that cities are able to implement policies that go beyond what is expected of them by their governments.

National governments continue to play an important role. Depending on the country, planning is guided by national laws that determine how often local governments are required to revise and update comprehensive development plans and the procedures that must be followed in their development. With the global demand for action on climate change, a shift occurs in many local governments that gives precedence to the environment over development interests. The Scottish Government, for instance, supports local governments by simplifying statutory processes around comprehensive plans, which enables local authorities to innovate with policies and strengthen protection of the environment.

When regulations and legislation are in support of local governments, planners and policy makers have the ability (and confidence) to go forward with plans and policies that may not be favourable in the short term but are beneficial in the long run. In relation to the physical built environment, building codes and regulations play an important role in the incorporation of green infrastructure and protection of nature. As the built environment changes slowly over time and buildings and other urban infrastructure are in use over several decades, the management of redevelopment and new buildings is critical. From a city perspective, regulation and policies that set a minimum amount of accessible green space for citizens and supporting instruments for achieving this are important. Next to such minimum standards, some cities have enacted special legislation. For instance, Copenhagen in Denmark became, in 2012, one of the first cities to have a mandatory green roof policy for municipal buildings as part of a wider climate change adaptation strategy (City of Copenhagen, 2012).

Strategies and Plans

For the majority of local governments, comprehensive development plans are the means by which they communicate their vision for their cities in a given period. Development plans are broad in their objectives. In the past, they were focused primarily on land-use and economic development. With the increased recognition that cities have the capacity to influence real changes that will help address climate change, these plans now include strategies for climate adaptation. An increasing number now include green infrastructure and the protection of nature. Some cities have even moved towards (or returned to) the inclusion of human health as a core objective.

Investments in green infrastructure can be effective at the building and neighbourhood scale, but also on a city and regional scale (see Box 4.2 for Barcelona, Spain). Local area plans permit planners and policy makers to engage directly with communities. This is an opportunity for building relationships with communities and informing the public on the benefits of nature protection and green infrastructure. Public consultation on neighbourhood plans provides an opportunity to show how improved land-use planning can be a cost-effective way to mitigate climate change and incorporate public health aspects (Younger et al., 2008).

In addition to action on developed land, protecting open, undeveloped spaces and interconnecting landscapes can equally contribute to providing micro-climatic regulation. Such green corridors and networks can provide further benefits such as connecting habitats and helping with protecting biodiversity on a landscape scale. This area of action is important from a city and regional perspective, and at times requires cooperation between different jurisdictions. The protected status of land according to national nature protection can facilitate the preservation of undeveloped land that

BOX 4.2: GREEN CORRIDORS FOR URBAN WELL-BEING: CITY OF BARCELONA, SPAIN

The City of Barcelona has developed an ambitious Green Infrastructure and Biodiversity Plan 2020 (Barcelona City Council, 2013). The plan builds on the idea of the multiple benefits of green space in the city and the creation of green corridors. This is achieved by maintaining existing green spaces and creating (small-scale) new ones. The Green Diagonal Avenue (Diagonal Verda) is an example of green connectivity in the city of Barcelona. It is a linear park across the eastern part of Barcelona that will connect a major highway interchange (Nus de la Trinitat) in northern Barcelona to the waterfront. To guarantee coherent development within the city, the Green Infrastructure and Biodiversity Plan is embedded in future city planning. To guarantee coherent development with the surrounding area of the city, a collaborative partnership with the Province (Diputacio) of Barcelona was established.

typically provides several ecosystem services simultaneously, including climatic regulation to moderate heat islands.

However, it should be mentioned that policy responses for addressing urban heat stress would need to go beyond aspects of the built environment. They also need to address the practical preparedness of health systems to manage extreme temperatures, raising awareness about good practices of self-protection during heatwaves (including emergency services such as cool rooms), and improve access to green spaces as well as social cohesion and trust-building (Huang et al., 2013). Thus, planning processes are not limited to the built environment but need to take a wider approach to identify vulnerable groups especially affected by heatwaves, including their geographic identification (Kunst and Britstra, 2013). Furthermore, a number of related policy areas such as emergency and disaster risk planning need to be involved.

Knowledge, Local Capacity and Cooperation

Analysing and mitigating the negative public health impacts of urban heat islands require knowledge and capacity. Regional and local governments are recognised for their potential to move the climate adaptation agenda forward (Corburn, 2009). National governments also have a role in supporting local governments. Countries such as France have developed health-warning systems for prolonged heatwaves (see Box 4.3 for Ghent in Belgium). Cities still have to plan responses to heatwaves for both the short and the long term. For the short term, during the course of a heatwave, the WHO has provided guidelines for developing a heat plan. For the long

BOX 4.3: GHENT, BELGIUM – CLIMATE ADAPTATION STUDY AND IMPLEMENTATION USING GREEN INFRASTRUCTURE

Ghent in Belgium has the ambition to be a climate-proof city by 2030. The city authorities in Ghent have carried out a detailed analysis of heat stress and UHI dynamics for the city, including climate projections. City temperatures in Ghent itself are generally 3 degrees warmer than the surroundings, increasing to 8 degrees on hot days. In contrast, city parks are at least 1.5 degrees cooler than the surrounding city. The initial heat stress study was turned into policy-relevant information, to inform climate adaptation for the short, medium (2014–2018) and long term (to 2030), including the Climate Plan 2030 Ghent (Stad Gent, 2014).

run, cities can address the built environment and vulnerable populations (Buscail et al., 2012).

Cooperation between cities, regions and researchers can be important in this regard. The New York City Regional Heat Island Initiative is one example of such a partnership between the state, the city and scientists that has resulted in a response plan that uses a combination of solutions (green roofs, tree planting, surface lightening) to address UHI effects (New York City, 2015). Another example from a European context is the EU-UHI project that facilitates cooperation between eight cities (Budapest, Ljubljana, Modena, Vienna, Padua, Prague, Stuttgart and Warsaw) for addressing UHI as a common issue and that provides cities with data that can inform their risk-prevention strategies (UHI Project, 2014).

Communities and Individuals: Grassroots Projects

Local communities and individuals can become instrumental stakeholders and decision makers in implementing nature-based solutions for mitigating urban heat. Many of the benefits of nature are linked to behaviour and rely on citizens who make active use of green spaces. Consequently, for some benefits, citizen action and grassroots initiatives can provide powerful and cost-effective tools when green spaces are threatened by alternative uses or when public finances are tight. However, facilitating communities to engage in managing their own local environments is a complex process.

Several cities have demonstrated success with 'grassroots' initiatives that resonate with communities, such as Arnhem in the Netherlands and Glasgow or Manchester in the UK (see Box 4.4). Arnhem's heat mapping exercise has been used as an opportunity to engage the community in understanding the impacts of heat. Glasgow's greening efforts, initiated by

BOX 4.4: MANCHESTER, UK – RED ROSE FOREST PROMOTING STREET TREES

The city of Manchester has adopted the benefits of green infrastructure in its climate change action plan *Manchester: A Certain Future* (Manchester City Council, 2014), as well as a forthcoming green infrastructure strategy (Manchester City Council, 2015). These documents explicitly refer to using natural spaces to mitigate urban heat islands and contribute to climate change adaptation.

The Red Rose Forest initiative is a community-orientated organisation in the Greater Manchester area that has been one of the key stakeholders and implementers of green infrastructure in the city (Red Rose Forest, 2014). Building on the UK Government's Big Tree Plant, which completed its aim to plant 1 million trees in the UK by early 2015 (Forestry Commission, 2015), Red Rose Forest continue to promote tree planting in Greater Manchester, citing the benefit of reduced summer temperatures through shading. Some of their projects include a competition providing grants of up to £20 000 for local street greening projects, tree audits of Manchester, and a tree planting campaign using funding from a local waste management company. The initiative has also published a guide for urban street tree planting (Red Rose Forest, 2014).

the private sector and with community participation, have gained a foothold in the city's current planning actions.

Generally, local governments are making progress developing and implementing policies that mitigate the impacts of climate change and adapt the urban environment. However, responses to the health impacts of climate change have been largely reactive and have only recently become more proactive as climate science shows that temperatures will continue to rise. As cities have demonstrated, formulating policies that enable people to mitigate and adapt to increased temperatures will require the acknowledgement of benefits and barriers, collaboration by all stakeholders, and education and research.

Synthesising the Role of Stakeholders, Policies and Measures

Table 4.1 synthesises a number of the above aspects that are relevant at the city, regional, country and international level. A wider use of carefully selected and strategically placed green infrastructure, together with a long-term policy for protecting nature in urban and suburban settings, can be an important element in mitigating urban heat islands and strengthening urban resilience with regard to climatic impacts. Understanding the challenge of urban heat stress, including its spatial dimension, and raising awareness of how to cope with it are important prerequisites to formulate strategies and plans that can target the short, medium and long term. A

Table 4.1 Climate resilience through nature governance: role of stakeholders, policies and measures: examples

	Knowledge – understanding the problem and solutions	Awareness and integration of knowledge	Policy, objectives, strategies and plans	Instruments, measures and legislation	Financing and investment
Global	WHO, IPCC and IPBES.	UNFCCC, UNCBD; SDGs and nature links.	UNFCCC targets (e.g. 1.5 °C Paris target); SDGs.	Conventions, protocols.	Climate adaptation funds (e.g. GEF).
EU	European Union research programmes (e.g. Horizon 2020), European Commission policy studies; mapping of ecosystem services (MAES).	Guidelines on for example Cohesion Policy for climate adaptation; case examples of nature-based solutions.	Climate strategies integrate nature; Cohesion Policy recognises climate adaptation; green infrastructure strategy recognises health and climate.	Supporting initiatives (e.g. Covenant of Mayors); sharing information and good practice; green city awards.	European Structural and Investment Funds; LIFE+ (ensure health link); TEN-GI (ensure health link).
National	National research, monitoring, modelling and mapping; national climate service centres.	Assessing the value of prevention (of for example hospitalisations, lost output).	Integration of nature in climate adaptation strategies and health strategies (and vice versa).	Clarifying of institutional responsibilities; inter-ministerial coordination; building codes (e.g. green space standards).	Environmental funds, grants, loans; green public procurement.
Regions and cities	Climate/heat/green infrastructure mapping; vulnerability assessments.	Independent evaluation of benefits to support action; awareness raising; heat emergency plans.	Integration of nature in urban development, climate adaptation, health and green infrastructure strategies.	Pilot projects with potential for scaling up; permitting citizens to plant pavements in front of their houses.	Investing in parks, tree-lined streets, green roofs on public buildings and incentivise citizen initiatives.

Table 4.1 (continued)

	Knowledge – understanding the problem and solutions	Awareness and integration of knowledge	Policy, objectives, strategies and plans	Instruments, measures and legislation	Financing and investment
Private sector	Assessment of cooling benefits of green roofs and walls; assessment of multiple benefits of landscape architecture and planning.	Communicate effectiveness of green solutions.	Opportunities for citizens to buy into reforestation or greening schemes.	Green private/sustainable procurement.	Building: investment in green roofs and green walls; landscape planners and landowners.
Civil society and citizens	Identifying vulnerable groups and needs of groups to make use of nature (i.e. access to use).	Communicate risks and opportunities for action and initiatives.	Collaborate for example with local or regional authorities in contributing to strategies and plans.	Tree ownership programmes, carbon offsetting schemes; bonds.	Own investment in green roofs; making use of public green space.
Research	Provide robust research on heat island reduction through nature and development of heat maps.	Expert groups supporting evidence-based policy making on nature/climate/health.	Research strategy on health–nature synergies.	Networks for expertise; field experiments; education and training.	National and European research grants; finance collaborative research centres.

Source: Own representation.

number of instruments exist, and cities and regions can make use of the existing legal framework to experiment with new approaches. For some aspects, changes to the legal framework might be useful, for example for defining minimum standards for green coverage and green spaces in the course of urban development. Finally, providing adequate financial resources for investing in green infrastructure and protected areas, as well as maintaining them, is an important condition for realising the health (and further benefits) of nature in urban and suburban areas.

CONCLUSIONS AND WAY FORWARD

The role of green spaces and wider green infrastructure, including protected areas, is likely to become more important in densely built European cities (and beyond) as a result of climate change. Green spaces not only provide micro-climatic cooling and regulation services, but can also serve as oases during heatwaves and provide health benefits to wider population groups. Solutions based on green infrastructure typically provide more than one benefit and in many cases multiple health benefits. A comprehensive assessment of these benefits is one of the challenges when developing green infrastructure solutions for urban heat islands. Without comprehensive inclusion of these multiple benefits, green spaces will invariably be undervalued.

The practical case examples from different European countries in this chapter point to the importance of local action for mitigating the UHI effects and the health impacts on the population. Thus strengthening the role of local communities, including their technical and organisational capacities, will help promote more proactive and collaborative action. This action can also test and evaluate different green infrastructure solutions, including different governance settings where health, social and nature stakeholders cooperate.

While local government can concentrate on developing programmes and plans, and encourage community engagement and collaboration across different groups, a role remains for the national and supra-national levels to catalyse local and regional action. The support can include providing information, such as scenarios of climate change in different regions, and guidance on the evaluation of the benefits and value of nature, as well as providing funding for projects that are of significance for achieving wider objectives. Finally, international organisations can support action at different levels by fostering an exchange of knowledge and by providing practical information such as the impacts of heatwaves on human health.

City greening is an important tool for adapting to the adverse effects of the urban heat island effect and climate change. However, reducing heat stress, which was the focus of this chapter, represents just one aspect contributing to urban well-being. Green infrastructure provides multiple health and social benefits, which should be considered as a multifarious approach to fostering urban resilience, which should be tailored to local contexts and the specific needs of citizens.

REFERENCES

Baccini, M., A. Biggeri, G. Accetta, T. Kosatsky, K. Katsouyanni, A. Analitis, H.R. Anderson, L. Bisanti et al. (2008), 'Heat effects on mortality in 15 European cities', *Epidemiology*, **19**(5), 711–719.

Barcelona City Council (2013), *Barcelona Green Infrastructure and Biodiversity Plan 2020*, Barcelona: Barcelona City Council.

Blue Green Dream (2015), 'Test and demo sites', http://bgd.org.uk/test-demo-sites/.

Bowler, D.E., L. Buyung-Ali, T.M. Knight and A.S. Pullin (2010), 'Urban greening to cool towns and cities: A systematic review of empirical evidence', *Landscape and Urban Planning*, **97**, 147–155.

Brink, P. ten, K. Mutafoglu, J.-P. Schweitzer, M. Kettunen, C. Twigger-Ross, J. Baker, Y. Kuipers, M. Emonts et al. (2016), *The Health and Social Benefits of Nature and Biodiversity Protection*, London: Institute for European Environmental Policy.

Buscail, C., E. Upegui and J.F. Viel (2012), 'Mapping heatwave health risk at the community level for public health action', *International Journal of Health Geographics*, **11**(1), 38–46.

Carter, T. and C. Butler (2008), 'Ecological impacts of replacing traditional roofs with green roofs in two urban areas', *Cities and the Environment*, **1**(2).

Ciscar, J.C., L. Feyen, C. Lavalle, A. Soria, F. Raes, M. Rozsai, A. Dosio, M. Donatelli et al. (2014), *Climate Impacts in Europe: Results from the JRC PESETA II Project*, JRC Scientific and Political Reports, EUR 26586EN, Luxembourg: Publications Office of the European Union.

City of Copenhagen (2012), *Green Roofs Copenhagen*, Copenhagen: Danish Ministry of the Environment and Danish Nature Agency – Danish Portal for Climate Change Adaptation, http://www.klimatilpasning.dk/media/631048/green_roofs_copenhagen.pdf.

Corburn, J. (2009), 'Cities, climate change and urban heat island mitigation: Localising global environmental science', *Urban Studies*, **46**(2), 413–427.

Depietri, Y., F.G. Renaud and G. Kallis (2011), 'Heat waves and floods in urban areas: Policy-oriented review of ecosystem services', *Sustainable Science*, **7**, 95–107.

EEA (European Environment Agency) (2012), *Urban Adaptation to Climate Change in Europe*, EEA Report 2/2012, Copenhagen: EEA.

EEA, JRC and WHO (2008), *Impacts of Europe's Changing Climate: 2008 Indicator-Based Assessment*, EEA Report No. 4/2008, Copenhagen: European Environment Agency.

Ennos, R. (2012), 'Quantifying the cooling benefits of urban trees', in Forestry Commission, *Trees, People, and the Built Environment: Proceedings of the*

Urban Trees Research Conference, http://www.forestry.gov.uk/pdf/FCRP017.pdf/$FILE/FCRP017.pdf.

Forestry Commission (2015), *The Big Tree Plant Funding Scheme*, http://www.forestry.gov.uk/england-bigtreeplant (accessed 27 May 2015).

Gill, S., J. Handley, A. Ennos and S. Pauleit (2007), 'Adapting cities for climate change: The role of the green infrastructure', *Built Environment*, **33**(1), 115–133.

Grand Lyon (2007a), *Aménagement et Eaux Pluviales*, Lyon, France: Grand Lyon.

Grand Lyon (2007b), *Climate Plan Greater Lyon: The Best Is Yet to Come*, Lyon, France: Grand Lyon.

Grand Lyon (2014), *Les Berges du Rhône*, http://www.lyon.fr/page/cadre-de-vie/ville-nature/les-berges-du-rhone.html (accessed 27 May 2015).

Hallegatte, S., N. Ranger, O. Mestre, P. Dumas, J. Corfee-Morlot, C. Herweijer and R. Muir Wood (2011), 'Assessing climate change impacts, sea level rise and storm surge risk in port cities: A case study on Copenhagen', *Climatic Change*, **104**, 113–137.

Handley, J. and J. Carter (2006), 'Adaptation strategies for climate change in the urban environment', ASCCUE draft final report to the National Steering Group, Centre for Urban and Regional Ecology, University of Manchester, UK.

Hathway, E.A. and S. Sharples (2012), 'The interaction of rivers and urban form in mitigating the urban heat island effect: A UK case study', *Building and Environment*, **58**, 14–22.

Huang, C., Adrian G. Barnett, Zhiwei Xu, Cordia Chu, Xiaoming Wang, Lyle R. Turner and Shilu Tong (2013), 'Managing the health effects of temperature in response to climate change: Challenges ahead', *Environment Health Perspective*, **121**(4), 415–419.

Hübler, M., G. Klepper and S. Peterson (2008), 'Costs of climate change – the effects of rising temperatures on health and productivity in Germany', *Ecological Economics*, **68**(1–2), 381–393.

IPCC (2014), *Fifth Assessment Report: Climate Change 2014: Synthesis Report Summary for Policy Makers*, Cambridge, UK: Cambridge University Press.

Jim, C.Y. (2013), 'Sustainable urban greening strategies for compact cities in developing and developed economies', *Urban Ecosystems*, **16**(4), 741–761.

Kovats, R.S. and K.L. Ebi (2006), 'Heatwaves and public health in Europe', *European Journal of Public Health*, **16**(6), 592–599.

Kovats, R.S., A. Haines, R. Stanwell-Smith, P. Martens, B. Menne and R. Bertollini (1999), 'Climate change and human health in Europe', *British Medical Journal*, **318**, 1682–1685.

Kunst, A.E. and R. Britstra (2013), 'Implementation evaluation of the Dutch national heat plan among long-term care institutions in Amsterdam: A cross-sectional study', *BMC Health Services Research*, **13**, 135–142.

MA (Millennium Ecosystem Assessment) (2005), *Ecosystems and Human Well-Being: Biodiversity Synthesis*, Washington, DC: World Resources Institute.

Manchester City Council (2014), *Manchester Green Infrastructure Strategy: Progress Update*, Manchester: Manchester City Council Neighbourhoods Scrutiny Committee.

Manchester City Council (2015), *Manchester Green and Blue Infrastructure Strategy*, http://www.manchester.gov.uk/downloads/download/6171/draft_manchesters_green_and_blue_infrastructure_strategy.

MMU (2011), '"Living" roofs take over city centre – two buildings at All Saints

planted', *Manchester Metropolitan University – News and Events*, http://www.mmu.ac.uk/news/news-items/1492/.
New York City (2015), 'Interactive tools and maps', New York City Health Department, http://www1.nyc.gov/site/doh/data/tools.page.
Oliveira, S., H. Andrade and T. Vaz (2011), 'The cooling effect of green spaces as a contribution to the mitigation of urban heat: A case study in Lisbon', *Building and Environment*, **46**, 2186–2194.
Peng, S., S. Piao, P. Ciais, P. Friedlingstein, C. Ottle, F.-M. Bréon, H. Nan, L. Zhou et al. (2012), 'Surface urban heat island across 419 global big cities', *Environmental Science and Technology*, **46**(2), 696–703.
Red Rose Forest (2006), 'Build your own Little Green Roof – type – wooden storage shed with a pitched roof', http://www.redroseforest.co.uk/web/images/stories/downloads/Quick%20guide%20-%20Little%20Green%20Roof%20on%20pitched%20shed.pdf.
Red Rose Forest (2014), 'Planting street trees', http://www.redroseforest.co.uk/web/images/stories/treespec2feb13pdf(2).pdf.
Renaud, V. and M. Rebetez (2009), 'Comparison between open-site and below-canopy climatic conditions in Switzerland during the exceptionally hot summer of 2003', *Agricultural and Forest Meteorology*, **149**(5), 873–880.
Rey, G., A. Fouillet, P. Bessemoulin, P. Frayssinet, A. Dufour, E. Jougla and D. Hémon (2009), 'Heat exposure and socio-economic vulnerability as synergistic factors in heat-wave-related mortality', *European Journal of Epidemiology*, **24**(9), 495–502.
Stad Gent (2014), *Ghent Climate Plan 2014–2019*, https://stad.gent/sites/default/files/page/documents/Ghent%20Climate%20Plan%202014-2019.pdf.
Susca, T., S.R. Gaffin and G.R. Dell'Osso (2011), 'Positive effects of vegetation: Urban heat island and green roofs', *Environmental Pollution*, **159**, 2119–2126.
TEEB (2011), *The Economics of Ecosystems and Biodiversity in National and International Policy Making*, ed. Patrick ten Brink, London: Earthscan.
Teuling, A.J., S.I. Seneviratne, R. Stöckli, M. Reichstein, E. Moors, P. Ciais, S. Luyssaert, B. van den Hurk et al. (2010), 'Contrasting response of European forest and grassland energy exchange to heatwaves', *Nature Geoscience*, **3**, 722–727.
Tzoulas, K., K. Korpela, S. Venn, V. Yli-Pelkonen, A. Kazmierczak, J. Niemela and P. James (2007), 'Promoting ecosystem and human health in urban areas using green infrastructure: A literature review', *Landscape and Urban Planning*, **81**, 167–178.
UHI Project (2014), 'Urban Heat Island Decision Support System', http://eu-uhi.eu/dss/.
Upmanis, H., I. Eliasson and S. Lindqvist (1998), 'The influence of green areas on nocturnal temperatures in a high latitude city (Göteborg, Sweden)', *International Journal of Climatology*, **18**(6), 681–700.
Vandentorren, S., F. Suzan, S. Medina, M. Pascal, A. Maulpoix, J.-C. Cohen and M. Ledrans (2004), 'Mortality in 13 French cities during the August 2003 heat wave', *American Journal of Public Health*, **94** (9), 1518–1520.
Watkins, R., J. Palmer and M. Kolokotroni (2007), 'Increased temperature and intensification of the urban heat island: Implications for human comfort and urban design', *Built Environment*, **33**(1), 85–96.
WHO (2005), *Ecosystems and Human Well-Being: Health Synthesis*, Report of the Millennium Ecosystem Assessment, http://www.millenniumassessment.org/documents/document.357.aspx.pdf.

Younger, M., H.R. Morrow-Almeida, S.M. Vindigni and A.L. Dannenberg (2008), 'The built environment, climate change, and health: Opportunities for co-benefits', *American Journal of Preventive Medicine*, **35**(5), 517–526.

Zoulia, I., M. Santamouris and A. Dimoudi (2009), 'Monitoring the effect of urban green areas on the heat island in Athens', *Environmental Monitoring Assessment*, **156**, 275–292.

Zupancic, T., C. Westmacott and M. Bulthuis (2015), *The Impact of Green Space on Heat and Air Pollution in Urban Communities: A Meta-narrative Systemic Review*, Vancouver, Canada: David Suzuki Foundation.

5. Coproducing resilience through understanding vulnerability

Vivek Shandas, Anandi van Diepen, Jackson Voelkel and Meenakshi Rao

> Coproduction helps to understand the full complexity of places and helps to broaden the scope of the possible answers. Space serves as a medium and as an integration frame for human activities. (Louis Albrechts, 2013: 56)

INTRODUCTION

We review a collaborative city–university vulnerability assessment from two perspectives: (1) an academic, phenomenological lens; and (2) a practical and empirical lens. We argue that, in order to understand resilience, we must first ask the question "Resilience to what and for whom?" To that end, we have characterized ways in which two major urban environmental hazards, namely air pollution and urban heat, have affected neighborhoods differentially in one city in the Pacific Northwest of the United States, Portland, Oregon. Empirically, we explore neighborhood-level exposure to air pollution and extreme heat, as well as these spatially defined communities' physiological sensitivity to the two hazards, and their social capacity to adapt to them. Phenomenologically, we argue that this collaboration is an effort in *coproduction*, encapsulating cooperation between scientific experts and governmental authorities in the production of knowledge toward a socially determined goal in the public interest.

In the Pacific Northwestern United States, the mid-size city of Portland is a longtime leader in municipal planning for climate action. In 1993, the city adopted one of the first climate action plans (CAP) worldwide, with subsequent updates driven largely by citywide imperatives for mitigation of greenhouse gases (GHGs). The newest plan (see www.portlandoregon.gov/bps/article/531984) contains ambitious goals of reducing GHG emissions to 40 percent of 1990 levels by 2030. Unlike many other such plans, this CAP emphasizes implementation of adaptation strategies, alongside mitigation, as a major plank in the plan's platform. This calls forth questions of which areas and which communities, in the city, are most in need of adaptation and preparedness—that is, questions of what factors create and reinforce vulnerability. Yet, given our limited understanding of climate vulnerability, and the paucity of empirical evidence on how municipal planning might foster and implement adaptation, responding to future shifts in climate regimes presents substantial challenges, such as: *Which communities and neighborhoods are most affected by climate hazards? How are human health and welfare at stake in such effects? How might transformed management systems respond to episodic events and persistent impacts? How can community-based and institutional adaptation complement one another, toward resilience?* These are general questions applicable to most if not all cities. Here we formulate responses to the informational needs of the city of Portland, particularly its climate action plan's directives for adaptation as a path to build resilience; we responded to such informational needs by developing a collaborative research-planning process that facilitated our engagement in scholarly conversations on the theoretical meaning and epistemological utility of vulnerability. The proactive nature of Portland's local efforts for climate change mitigation and adaptation, as several prior scholars have affirmed (Rutland and Aylett, 2008; Aylett, 2013), distinguishes the city as a promising case study for addressing these questions. In addition, the project allowed us to involve our planner partners in these conversations, recursively testing the usefulness of coproduction in vivo.

In this chapter we review the findings of four collaborative workshops of neighborhood-scale vulnerability, in which planners and researchers engaged with and coproduced spatial and statistical analyses of such vulnerability; moreover, we posit a conceptual framework underlying the project's "collaborative rationality" (Innes and Booher, 2010). Such a framework, shared between scholars and practitioners, serves as an epistemological common ground for building deliberative and informed efforts to identify vulnerability and build resilience to environmental, and in particular climate, hazards.

CLIMATE VULNERABILITY IN THE CITY: SOCIALLY CONSTRUCTED ADAPTIVE LANDSCAPES

Our argument for coproduction as a model for urban resilience rests on a recent collaborative project between planners and researchers. The Portland Climate Action Collaborative consists of two primary partners, the city of Portland's Bureau of Planning and Sustainability and Portland State University. Other stakeholders, including representatives from regional public authorities that serve on health and environmental quality programs, including environmental justice organizations, joined the planners. The goal was to generate knowledge informing planning interventions for the city's adaptation to two major climate change-related hazards: (1) extreme heat, particularly concentrated in urban heat islands; and (2) air pollution from mobile sources, that is, vehicular traffic. These two hazards were chosen because of their explicit mention in the most recent climate action plan, and for their immediate and acute implications for human health.

The university research team, along with the above practitioners, formed a task force that met for four in-depth workshops, jointly facilitated by the researcher and planner leading the team, with the goal of assessing vulnerability as a first stage of building regional resilience. The assessment was aimed at developing shared knowledge of Portland's geographic and demographic concentration of need for adaptive intervention, based on the two hazards. The short-term theory of change is that planners' collaboration with knowledge producers and knowledge implementers could inform adaptation, effectively targeting the most vulnerable communities in Portland, thus supporting an agenda of equity planning (*sensu* Krumholz and Forester, 2011), which pervades the culture of planning in this city. The long-term hypothesis, as we establish later in this chapter, is that such targeted adaptation—the coproductive result of "making sense together" (Healey, 1992: 148) through communicative planning and adaptive comanagement of urban ecosystems—will build institutional and tangible infrastructure for sustained resilience of communities erstwhile vulnerable to the two hazards.

In describing the process of coproduction of urban resilience, this chapter links theoretical scholarship on collaborative governance, particularly in planning, to empirical, geospatial measurement of vulnerability, to which we contribute with our study of Portland's socio-ecological landscapes by way of vulnerability's dynamics in "socially constructed and adaptive landscapes" (McLaughlin and Dietz, 2008). Elsewhere in much of the resilience literature, vulnerability is taken for granted more generically as a subject's risk of or susceptibility to suffering serious impacts from a hazard (Adger, 2006; Cutter et al., 2006). Other studies measure access to or lack of resources, including "entitlements" (Watts and Bohle, 1993 and Ribot, 1995, following Sen, 1981), to cope with existing political economic structures, particularly poverty, that create and perpetuate vulnerability to environmental hazards. A plethora of earlier studies uses multi-dimensional data on social and environmental indicators of risks, hazards and their compound effects, to formulate a composite, sometimes predictive, measure of vulnerability (Kasperson and Archer, 2005; Birkmann, 2007; Collins et al., 2009; see also Fothergill, 1996). Variables that the literature establishes as contributing to a composite definition of vulnerability include socio-economic status, demographic factors of race, ethnicity, gender and age, and factors of the social and built environment, such as housing quality (Younger et al., 2008; Uejio et al., 2011). We identify these variables as socio-ecological risk factors—conditions that link changes to the global environment with ramifications to human health—within the nexus of the "socially constructed adaptive landscape." McLaughlin and Dietz (2008: 107) describe this landscape as "a three-dimensional surface."

> [T]he X and Y axes represent the structural, behavioral, and cultural characteristics of different roles, routines, households or organizations and the [third] axis represents the fitness of different combinations of those characteristics. The contours of the landscape . . . are dynamic and shift over time in response to changes in the underlying biophysical, social, economic, political, and discursive environments. (McLaughlin and Dietz, 2008: 107)

For our vulnerability assessment, this model of urban spatial dynamics between human behavior and health, on the one hand, and environmental conditions, on the other, serves as a comprehensive unit of analysis. Working in the realm of governance, and aware that political economic conditions structure human vulnerability as much as environmental conditions do, we concur with the conception of power embedded in this landscape model. "Social power within this framework implies the ability to reshape the contours of the landscape and thereby alter the founding, disbanding and change of social forms at various levels of analysis" (McLaughlin and Dietz, 2008: 107).

Thus this vulnerability assessment is an effort to develop "social power," of planning governance as well as of citizens and communities, in adapting Portland's landscapes toward resilience in human health and well-being to extreme heat and air pollution. Indeed the social construction and adaptation of landscape processes have always been underway. Yet here the directive that the vulnerability assessment informs is intentionally to construct urban landscapes facilitating adaptation, and thus raising quality of life for the most vulnerable, even as the city and world confront climate change. Coproduction, the process of generating knowledge the vulnerability assessment invokes, is likewise always underway. Again the effort here is to construct not only landscapes, but also understandings of landscapes that mend the social inequities that have led to vulnerability and disparities therein across space. We ultimately intend this inquiry to lead to a grounded understanding of how efforts toward resilience might proceed. We turn now to describing the theory and methodology underlying the project's first phase.

CHARACTERIZING VULNERABILITY

The hazard-specific vulnerability assessment we performed reveals differential levels of exposure, sensitivity and adaptive capacity to extreme heat and degraded air across neighborhoods, operationalized as census block groups (CBGs), in the city of Portland, with the city construed as a mosaic of "socially constructed adaptive landscapes" nested at different scales (McLaughlin and Dietz, 2008). Our integrated and applied methodological approach to assessing vulnerability rests on a theoretical explanation of the systemic causality underlying vulnerability, variegated across inequitable urban space. Vulnerability is a property emerging from the cumulative effects of three measurable risk factors: *high exposure*, *low sensitivity* and *low adaptive capacity*, all in relation to a given hazard in specified socio-ecological urban space (Figure 5.1). Thus our study has characterized Portland neighborhoods by these three dimensions as indicators of vulnerability itself. As a whole, the interdisciplinary literature on vulnerability to climate change, and to environmental hazards broadly, lays the conceptual and empirical groundwork for studying patterns of morbidity in the occurrence of the three phenomena indicating vulnerability, as they occur in specified socio-ecological space. The conceptual underpinnings of this chapter's account of our vulnerability assessment form the base of long-term scholarly conversations on how collaboration and coproduction might help to reduce those vulnerabilities, and hence bring forth resilience.

Undergirding the assessment we have conducted, we conceptually frame vulnerability as a phenomenon occurring within socio-ecological systems. Our understanding of vulnerability stems from a "critical realist epistemology" (McLaughlin and Dietz, 2008) in which the measurable unit of analysis is the "socially constructed adaptive landscape." To define vulnerability, it is necessary to deconstruct the concept into events and events into phenomena. To that end, we take an "event ecology" (Vayda and Walters, 1999) approach, beginning with an environmental event, a heat wave, for example, which is a spatially and temporally specific manifestation of a climate hazard. *Climate vulnerability, as a phenomenon, entails reinforcing feedback; a subject can be considered vulnerable when the effects of the hazardous event to which he or she is susceptible actually worsen health and welfare in the short and long term, thus compounding that vulnerability in advance of the next event.* For example, on a day with inordinately high air pollution a child might suffer a very traumatic asthma attack, increasing the severity and risk of an attack on the next poor-air day. Additionally the financial burden of responding to the attack could diminish the family's capacity to cope in the event of another attack. According to differential socio-ecological conditions—including vegetation, individual health,

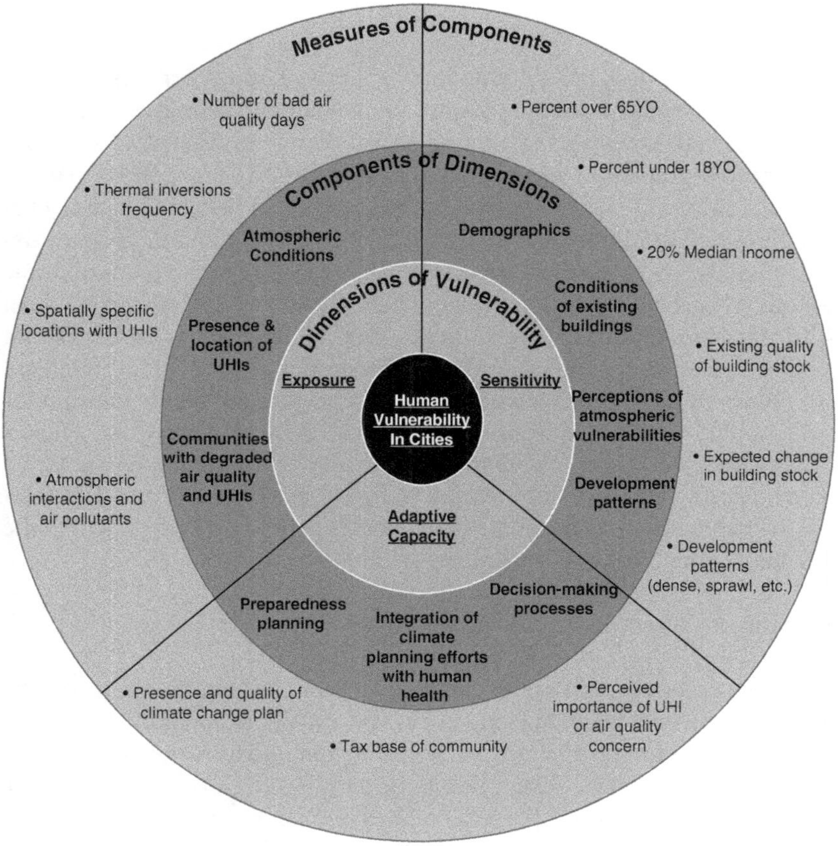

Source: Adapted from Polsky et al. (2007).

Figure 5.1 Vulnerability scoping diagram (VSD) concept diagram

housing quality and other factors of the built environment—the event will exact health impacts unequally across the urban landscape. A heat wave, for example, is more likely to cause illness and death in individuals within impoverished, elderly and socially isolated subpopulations, relative to affluent, middle-aged and socially integrated individuals (Reid et al., 2012). This example of a suite of socio-ecological factors describing the "adaptive landscape" (McLaughlin and Dietz, 2008) is useful for empirical description of that landscape, via the commonly measured conditions surrounding vulnerability. However, the definition of vulnerability, as a whole, is more than the composite sum of its commonly measured parameters.

CASE STUDY

We set out to assess human vulnerability to heat stress and degraded air quality as a means of building toward municipal-level resilience to extreme climate events; this involved engaging local planners, other decision-makers, and researchers in a coproduction of knowledge model. By pairing a local university with regional stakeholders, we examine conceptual descriptions of vulnerability and the extent to which human communities are impacted by urban heat stress and degraded air quality from multiple perspectives. Few studies combine a collaborative, coproductive model with data-intensive mapping exercises that forecast the social, environmental and economic impacts of alternative spatial options (although see Abunnasr et al., 2015). By assessing the three measurable dimensions of vulnerability—exposure, sensitivity and adaptive capacity—we engage local stakeholders in characterizing Portland neighborhoods in terms of climate-induced stressors related to urban heat, urban form and land use stressors specific to air quality, interaction of the two, and specific "hotspot" neighborhoods experiencing both.

Preparation: Operationalizing Vulnerability

Planning and scientific communities are both intimately aware of the human impacts of climate change. Yet it is critical to integrate these perspectives and inquire how planners, in remaking adaptive urban environments, might cooperatively apply emergent knowledge from the sciences of climate change. With this project, we explain one instance of "joined-up" (Agyeman et al., 2003) research and planning at the intersection of these two epistemic communities. As a means of engagement and knowledge discovery between the two communities, coproduction's proximate goal, as we see it, is to expand each epistemic community's knowledge base; the ultimate goal is to improve the two groups' effectiveness in protecting human health and well-being in the face of climate change.

As we conducted it, the vulnerability assessment, as a coproductive, joined-up process, can aptly be represented in terms of both research and planning. Central to this assessment, a series of four workshops—held with parties from several City bureaus, the State of Oregon's Health Authority, Multnomah County Department of Public Health, and several non-governmental organizations—shaped how the team assessed vulnerability. The stakeholder process consisted of four workshops of approximately 25 attendees, allowing individuals to address specific questions about the construction of this project. *Coproduction of knowledge involved three distinct steps: (1) understanding the concept of vulnerability through the vulnerability scoping diagram; (2) identifying factors that contribute to hazard-specific vulnerability and the interaction of factors that amplify vulnerability; and (3) developing an online interactive vulnerability assessment tool to identify the areas most resilient to these environmental hazards.*

The workshops, which occurred through the 2014–2015 academic year, used the VSD (Figure 5.1) as a means for synthesizing the multiple dimensions of vulnerability, and as a heuristic for discussion. Discussion of participants' diverse understandings of vulnerability, including its meanings and manifestations, enabled us to identify socio-demographic variables that support an analytical process for rendering vulnerability across the city landscape. Specifically, the workshops brought out participants' knowledge and perspectives of the conditions (e.g. age) identifying subpopulations with both high sensitivity to climate hazards and low adaptive capacity to manage their exposure to these environmental pressures. Participants identified the individual, communities, and built environment factors that could be directly linked to the VSD. As a result, the research team could then work with workshop participants to procure and curate the necessary datasets to establish a working definition of vulnerability.

Thus the workshops resulted in a set of coproduced socio-demographic variables (Table 5.1) to identify the sensitivity and adaptive capacity dimensions of vulnerability to the two ecological hazards of air pollution and urban heat. Data for each of these variables were procured from several data sources, including United States Census Bureau American Community Survey (ACS) 2013 block group data, Oregon Department of Motor Vehicles via Oregon Health Administration, Multnomah County Tax Assessor Database via Portland Bureau of Planning and Sustainability (BPS), and Voter Activation Network via Coalition for a Livable Future (CLF).

The exposure dimension for air pollution and urban heat was constructed from observationally based, highly spatially resolved models of summer-time nitrogen dioxide (Rao et al., 2014) and urban heat (Hart and Sailor, 2009) in the Portland Metro Area. Exposure, sensitivity and adaptive capacity metrics were then combined, as described below, to create a single "total vulnerability."

We note that both exposure variables are based in spatially explicit descriptions and are in the form of a grid-based data model. As a result, extra steps were taken before integration of the tabular and vector formats, which was the structure of the socio-demographic data. In order to integrate the socio-demographic and exposure variables, the surfaces of urban heat and air quality first had to be converted into census block group geographies. This was accomplished by calculating the mean value of the pixels for each environmental variable within each CBG in Portland. This mean value was in turn appended onto existing CBG data for tabular comparisons and

Table 5.1 Data and sources

Hazard		Variable	Description	Source	Year
Urban heat	Air pollution				
✔	✔	Adults over 65	Adults older than 65 as a percentage of the total block group population.	American Community Survey	2013 (2009–2013)
✔	✔	Youth under 18	Minors under 18 years old as a percentage of the total block group population.	American Community Survey	2013 (2009–2013)
✔	✔	Persons below 50% poverty line	Percentage of the block group population who are below 50% of the poverty line*.	American Community Survey	2013 (2009–2013)
✔	✔	Adults without a high school degree	Percentage of adults over 25 years old in the block group who do not have a high school degree or equivalency.	American Community Survey	2013 (2009–2013)
✔	✔	Communities of color	All non-white individuals as a percentage of the total block group population.	American Community Survey	2013 (2009–2013)
✔	✔	BMI	Block group averages of BMI scores. (BMI = mass(kg)/height(m^2))	Oregon State Department of Motor Vehicles	2012
✔		In-home air conditioning	Percentage of residential buildings with air conditioning per block group (excluding window units).	County Tax Assessor Database	2015

✔	✔	Language status other than English as primary language	Percentage of the block group population who do not speak English as their first language and have poor (or no) English speaking abilities.	American Community Survey	2013 (2009–2013)
✔	✔	Adults over 65 living alone	Adults older than 65 who live by themselves as a percentage of the total block group population.	American Community Survey	2013 (2009–2013)
	✔	Voter turnout	Percentage of the block group population who voted in the three most recent general elections.	Voter Activation Network	2012

manipulation further into the workflow. A simple scoring scheme was used to score the CBGs for each socio-ecological variable. A CBG was given a score of 1 if it fell into the worst quintile of the socio-ecological variable; otherwise it was given a score of 0. The scores for all socio-ecological variables for each CBG were summed to create the "total vulnerability" score. In essence, the "total vulnerability" score represents the number of socio-ecological variables for which the CBG lies in the worst quintile. It should be noted that giving a score of 1 to all variables that fall in the worst quintile equally weights their importance in the final score. Weights can be readjusted, as dictated by knowledge and circumstance.

The identification of neighborhood-scale stressors—for example, urban heat islands within socially vulnerable areas—enables stakeholders to gain leverage for adaptive strategies in the short and long term. Depending on the weighting of vulnerability factors, we observe differences in the areas where the factors intersect. We found that, if we weight the pre-existing health condition as a more important factor in determining vulnerability, we arrive at a different set of results. To that end, and for the sake of discussion, in this initial stage we weighed all the factors evenly. This model can and should be specified to reflect contextual differences in factors affecting vulnerability in "socially constructed adaptive landscapes," yet ours is a first pass at operationalizing the construct of vulnerability based on input from multiple stakeholders.

By combining the exposure (environmental), sensitivity and adaptive capacity (socio-demographic) factors, we observe a distinct set of patterns that are consistent with the understanding of stakeholders involved in this project (Figure 5.2).

The combined set of factors contributing to vulnerability across the city illustrates variation and clustering. Areas toward the east of the city, for example, are darker, which represents a convergence of multiple factors: high exposure and sensitivity to urban heat stress and degraded air quality, and low adaptive capacity. The stakeholders consistently remarked that the resulting maps were not surprising given the city's history of poverty and disinvestment in the eastern parts of the city. Other areas include the city's Chinatown, which is the oldest area in the central business district, neighborhoods adjacent to freeways, and areas of North Portland, much of which sits adjacent to the region's largest interstate freeway.

Taken together, the outcomes of the workshops were instrumental in stakeholders delineating hazard-specific vulnerability as a case-by-case condition as a result of the complex feedback among interacting factors. For example, one stakeholder described a cascade of interacting factors of vulnerability, such that a pre-existing health condition based on genetic disposition may reduce an individual's ability to perform well at school,

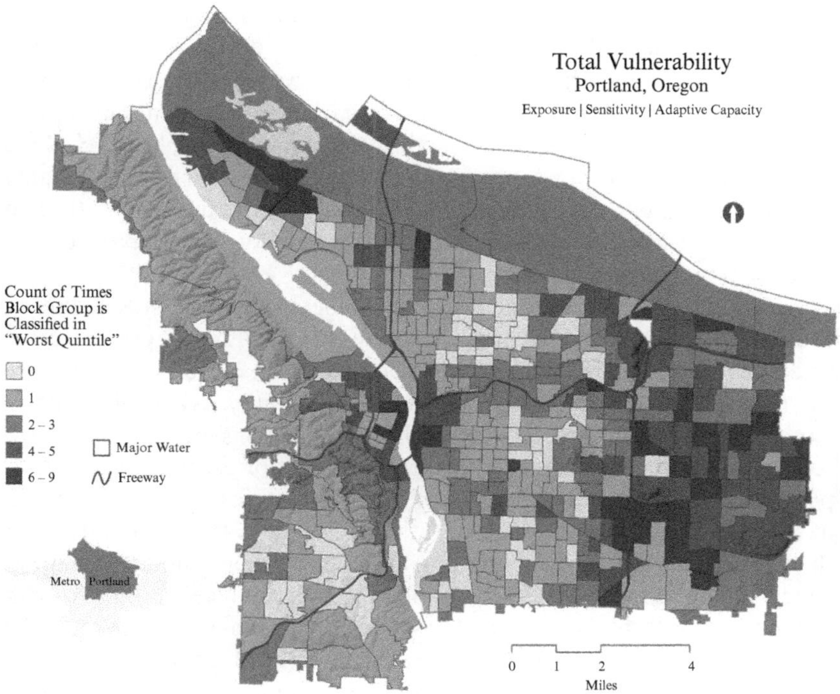

Note: Darker colors reflect greater vulnerability as defined by the combination of air quality, urban heat island effect, and socio-demographics identified in Table 5.1.

Figure 5.2 Final map of vulnerable populations (by census block group)

thereby limiting the individual's educational advancement, professional networks, and income, and ultimately reinforcing vulnerability as discussed above. In addition, the workshops helped to acknowledge that different stakeholders, based on their professional obligations and objectives, were given to emphasize different aspects of vulnerability (e.g. health for the county public health stakeholders, physical design for urban designers, etc.). These insights, while not evidenced in the outputs of the workshop, nevertheless help to underscore the social constructedness of landscapes marked by disparate vulnerability, requiring at once an understanding of the factors at play, those who are disproportionately affected, and what to do about it.

Coproducing Vulnerability Assessments

Under the rubric of vulnerability as a set of socio-ecological conditions that enable the mutually reinforcing harms of a hazard, the vulnerability scoping diagram above identifies the three dimensions of vulnerability, what socio-ecological conditions indicate those dimensions, and what metrics might characterize those conditions. As we show, these are measurable constructs, yet as individual or sets of variables they do not phenomenologically evince *vulnerability* as a construct. Importantly, however, these are the preconditions our coproductive process has shown to be associated with vulnerability's mutually reinforcing functioning, and are critical to scoping—that is, assessing—the exposure, sensitivity and adaptive capacity of socio-ecological systems.

We, as researchers, and our primary partners, environmental planners, collaboratively designed this empirical research with the purpose of characterizing differential socio-ecological landscapes in our city by their relative risk to the harmful feedback effects—vulnerability, as described above. Crucial to our coproductive research process was synthesizing known socio-ecological risk factors to the health impacts of extreme heat and air pollution—established in pertinent scientific literatures—with the environmental and socio-demographic conditions relevant to Portland and within the purview of the city's planning community. Where climate change is concerned, quantitative characterization of the human geographies of risk is a method well substantiated across a multidisciplinary spectrum of the sciences. Indeed, in a review paper, de Sherbinin (2014) provides a synopsis of research applying multi-layered spatial and numerical data to represent climate hazards' relative risk of harming human health and well-being. A process he has termed "climate change hotspots mapping" has measured the incidence of vulnerability across space as a composite function of its three sub-dimensions: exposure, sensitivity and adaptive capacity. De Sherbinin calls this dimension of the concept "outcome vulnerability" (2014: 24), contrasting it with "context vulnerability . . . [that is,] the intrinsic characteristics of a system . . . rooted in political economy" (2014: 24). Because it neatly deconstructs the concept into measurable dimensions of climate impact (*exposure*) and human response (*sensitivity* and *adaptive capacity*), outcome vulnerability is amenable to hotspot mapping—that is, the combination of multiple datasets to yield an aggregate measure of vulnerability. In contrast, context vulnerability accords with our own conceptualization of vulnerability as a phenomenon marked by reinforcing feedback. Meanwhile outcome vulnerability accords with the idea of differential socio-ecological conditions, which our coproductive project identified as indicators of exposure, sensitivity and adaptive capacity.

We believe that our project well served the informational needs of our collaborators in planning. Decision-makers and implementers of Portland's planning of the ecological and built environments need city-scale understandings of relative risk—stylized as intra-city vulnerability—to hazards, for appropriate focusing of their interventions to build preparedness by reducing such risks. Moreover, integration of multiple data layers at this relevant scale is a research design amenable to collaborative aggregation of variables germane to the capacity of our community partners. As de Sherbinin points out, maps of hotspots often "help to communicate . . . in a manner . . . easier to interpret than text" (2014: 23). Provided there is some degree of congruity or complementarity in researchers and "decision-makers' mental models" (2014: 34), the collaborative epistemological approach we call coproduction in vulnerability assessment may to an extent "rely . . . on expert opinion" in defining research questions and geographic extent. While composite, aggregate measures of vulnerability render visually intelligible models of a myriad of risk factors, de Sherbinin's review suggests that these methods of characterizing vulnerability are subject to various evaluations of internal validity. While data on the differential occurrence and severity of climatic events in space directly describe the dimension of exposure, for data sensitivity and adaptive capacity, "close surrogates" (de Sherbinin, 2014: 11)—e.g. income and education respectively—are invoked. As we discuss above, research "incorporating conceptual understanding[s of] political and social marginalization, gendered relationships, and physiological differences" (de Sherbinin, 2014: 11, quoting Kasperson and Archer, 2005) requires an "event ecology" approach: specified inquiry into how the multi-layered socio-ecological context relates to climatic events' outcomes to human health.

The collaborative design of this project, as well as its empirical results to date, situated the team of scientists and practitioners in a hybrid epistemic phase that was at once a problem space as well as a decision space. The next phase of the venture, on one hand, will involve the team's use of these findings on differential intra-urban vulnerability to identify *problems* that planners can address as well as problems that researchers can address. On the other hand, the next phase will be to identify *decisions* that planners can make as well as decisions that researchers can address. The collaborative vulnerability assessment here described equips planners to diagnose problematic features of socio-ecological landscapes in the city, and address those that are remediable. It equips us researchers to address epistemological problems, such as threats to internal validity and the ecological fallacy, by designing a subsequent methodology (in Phase 2) that will triangulate our analysis of human vulnerability in the hitherto identified adaptive landscapes in Portland, sampling from neighborhoods to collect qualitative data at more granular scales. Problem identification, then, from the planners' perspective, will therefore inform their decisions on the construction and implementation of urban design and development. From the researchers' perspective, problem identification will yield a more comprehensive and validated dataset on the causes and consequences of vulnerability to climate hazards.

CONCLUSIONS

The goals of our university–city research partnership have been to generate a body of evidence on Portland's existing vulnerability to climate change hazards, to develop an actionable understanding of resilience, and ultimately to coproduce interventions toward hazard-specific resilience in neighborhoods. The core methodology of the project's first phase, as reported here, was the spatial statistical scoping of the socio-ecological landscape of vulnerability in Portland, an assessment whose design was coproduced in a collaborative forum with stakeholders in the city's planning community. Although the results of our assessment and the chapter itself focus on Portland, the process we have developed for engaging researchers and planners is generally applicable. This experiment in scientifically informed coproduction of planning knowledge provides a general, if preliminary, model in collaborative comanagement of urban environmental quality. Moreover and more specifically, this project has generated a set of lessons on the value of this practice and its cobenefits to scientific and planning progress. Incorporating participating planners' and other stakeholders' understandings of the dimensions of vulnerability, for example, informed the variable selection step of research design. In turn, on the researchers' side, the statistical modeling procedure described above produced maps of the geographic extent of vulnerability—conceptualized collaboratively—across the city. Coupled in this way, the results of the vulnerability assessment will then work to feed the future of adaptation and resilience in Portland.

An ultimate outcome of this collaboration is to usher in an adaptive future wherein human well-being in cities is not only preserved, but also improved with increasing social equity, in the face of intensifying climate hazards. A practical and conceptual turn that portends this development is a "monitoring and phasing" (Abunnasr et al., 2015: 139) approach to adaptation planning. An example of such an approach includes a recent proposition that Abunnasr and colleagues argued, which suggests that adaptation planning should deploy mechanisms of monitoring and phasing in order to overcome barriers of temporal uncertainty in climate change forecasting. They stress that adaptive interventions, that is, those that can be revisited over time, offer cobenefits in that they serve as heuristics with which planners may "test for maladaptation" (Abunnasr et al., 2015: 139). Maladaptation, they explain, arises from the unintended consequences of adaptive strategies that may engender new amplifying or reinforcing dynamics (that is, vulnerabilities) even as they meet an original objective. A common maladaptation to climate change is the expansion of air conditioning, increasing the release of GHGs, and actually worsening the problem itself. We further contend that a coproductive process that engages multiple sectors to define, evaluate and address the problem will lead to fewer instances of maladaptation, and deepen opportunities for creating climate resilient communities.

Resilience, as a set of socio-ecological conditions, is an end goal that will marry mitigation and adaptation and, further, will balance inter- and intra-generational equity by resisting maladaptations that "disproportionately burden the most vulnerable" or "limit the choices available to future generations" (Abunnasr et al., 2015: 139). With our team's vulnerability assessment, we argue that promoting long-term resilience in adaptive urban landscapes requires an equity mandate, given that issues of social justice, including racism, poverty, and health disparities, are climate justice issues as well. Resilience is more than merely a lack of vulnerability in a given urban landscape; it requires transformation. Adapting landscapes toward resilience requires but is not limited to repairing the inequities that cause the vulnerability we assessed. It is necessary, also, to engage in multilevel transformation to build what Tyler and Moench (2012) call an "urban climate resilience framework." Resilience requires "absorption of disturbances while maintaining structure and function" (Tyler and Moench, 2012) at the infrastructural, institutional and human-agent scales; responsiveness and learning in regard to hazards, which our assessment entails, are a major element of resilience at all three scales. Resilience also involves improving the existing structure and function of human landscapes where quality of life is low and disparities are deep. While resilience is a tall order indeed, we believe that the unequivocal severity of the hazards of climate change requires of scientists and planners collaborative, indeed coproductive, effort that is both proactive and ambitious.

ACKNOWLEDGMENTS

We thank the entire research and planning team and others who have made this work possible, including Michele Crim and Michelle Kunec-North (Bureau of Planning and Sustainability), Kari Lyons (Multnomah County Health Department), Sarah Armitage (Oregon Department of Environmental Quality), Betsy Clapp (Multnomah County Health Department), Brendon Haggerty (Oregon Health Authority) and Steve White (Oregon Public Health Institute). Funding came from the Bullitt

Foundation and the Institute for Sustainable Solutions at Portland State University.

REFERENCES

Abunnasr, Y., E.M. Hamin and E. Brabec (2015), 'Windows of opportunity: Addressing climate uncertainty through adaptation plan implementation', *Journal of Environmental Planning and Management*, **58** (1), 135–155.

Adger, W.N. (2006), 'Vulnerability', *Global Environmental Change*, **16** (3), 268–281.

Agyeman, J., R.D. Bullard and B. Evans (2003), 'Joined-up thinking: Bringing together sustainability, environmental justice, and equity', in *Just Sustainabilities: Development in an Unequal World*, Cambridge, MA: MIT Press, pp. 1–16.

Albrechts, L. (2013), 'Reframing strategic spatial planning by using a coproduction perspective', *Planning Theory*, **12** (1), 46–63.

Aylett, A. (2013), 'The socio-institutional dynamics of urban climate governance: A comparative analysis of innovation and change in Durban (KZN, South Africa) and Portland (OR, USA)', *Urban Studies*, **50**, 1386–1402.

Birkmann, J. (2007), 'Risk and vulnerability indicators at different scales: Applicability, usefulness and policy implications', *Environmental Hazards*, **7** (1), 20–31.

Collins, T.W., S.E. Grineski and M.D.L.R. Aguilar (2009), 'Vulnerability to environmental hazards in the Ciudad Juárez (Mexico)–El Paso (USA) metropolis: A model for spatial risk assessment in transnational context', *Applied Geography*, **29** (3), 448–461.

Cutter, S., B.J. Boruff and W.L. Shirley (2006), 'Social vulnerability to environmental hazards', in S.L. Cutter (ed.), *Hazards, Vulnerability, and Environmental Justice*, London: Earthscan, pp. 115–132.

Fothergill, A. (1996), 'Gender, risk, and disaster', *International Journal of Mass Emergencies and Disasters*, **14** (1), 33–56.

Hart, M.A. and D.J. Sailor (2009), 'Quantifying the influence of land-use and surface characteristics on spatial variability in the urban heat island', *Theoretical and Applied Climatology*, **95** (3–4), 397–406.

Healey, P. (1992), 'Planning through debate: The communicative turn in planning theory', *Town Planning Review*, **63** (2), 143–162.

Innes, J.E. and D.E. Booher (2010), *Planning with Complexity: An Introduction to Collaborative Rationality for Public Policy*, Abingdon, UK: Routledge.

Kasperson, R.E. and E.R. Archer (2005), 'Vulnerable peoples and places', in R. Hassan, R. Scholes and N. Ash (eds), *Ecosystems and Human Well-Being: Current State and Trends: Findings of the Condition and Trends Working Group*, Vol. 1, Washington, DC: Island Press, pp. 143–162.

Krumholz, N. and J. Forester (2011), *Making Equity Planning Work: Leadership in the Public Sector*, Philadelphia, PA: Temple University Press.

McLaughlin, P. and T. Dietz (2008), 'Structure, agency and environment: Toward an integrated perspective on vulnerability', *Global Environmental Change*, **18** (1), 99–111.

Polsky, C., R. Neff and B. Yarnal (2007), 'Building comparable global change vulnerability assessments: The vulnerability scoping diagram', *Global Environmental Change*, **17** (3), 472–485.

Rao, M., L.A. George, T.N. Rosenstiel, V. Shandas and A. Dinno (2014), 'Assessing the relationship among urban trees, nitrogen dioxide, and respiratory health', *Environmental Pollution*, **194**, 96–104.
Reid, C.E., J.K. Mann, R. Alfasso, P.B. English, G.C. King, R.A. Lincoln, H.G. Margolis, D.J. Rubado et al. (2012), 'Evaluation of a heat vulnerability index on abnormally hot days: An environmental public health tracking study', *Environmental Health Perspective*, **120**, 715–720.
Ribot, J.C. (1995), 'The causal structure of vulnerability: Its application to climate impact analysis', *GeoJournal*, **35** (2), 119–122.
Rutland, T. and A. Aylett (2008), 'The work of policy: Actor networks, governmentality, and local action on climate change in Portland, Oregon', *Environment and Planning D*, **26**, 627–646.
Sen, A. (1981), *Poverty and Famines: An Essay on Entitlement and Deprivation*, Oxford, UK: Oxford University Press.
Sherbinin, A. de (2014), 'Climate change hotspots mapping: What have we learned?', *Climatic Change*, **123** (1), 23–37.
Tyler, S. and M. Moench (2012), 'A framework for urban climate resilience', *Climate and Development*, **4** (4), 311–326.
Uejio, C.K., O.V. Wilhelmi, J.S. Golden, D.M. Mills, S.P. Gulino and J.P. Samenow (2011), 'Intra-urban societal vulnerability to extreme heat: The role of heat exposure and the built environment, socioeconomics, and neighborhood stability', *Health and Place*, **17** (2), 498–507.
Vayda, A.P. and B.B. Walters (1999), 'Against political ecology', *Human Ecology*, **27** (1), 167–179.
Watts, M.J. and H.G. Bohle (1993), 'Hunger, famine and the space of vulnerability', *GeoJournal*, **30** (2), 117–125.
Younger, M., H.R. Morrow-Almeida, S.M. Vindigni and A.L. Dannenberg (2008), 'The built environment, climate change, and health: Opportunities for co-benefits', *American Journal of Preventive Medicine*, **35** (5), 517–526.

6. Climate resilience and sustainable development: challenges and options for small island developing states

Ramón Pichs-Madruga

INTRODUCTION

The international debate on development in the last four decades has incorporated environmental sustainability as a key ingredient. The concept of sustainable development as defined by the Brundtland Report of 1987 and the Rio Declaration on Environment and Development of 1992 refers to three main dimensions: economic, social and environmental pillars, as basic ingredients of the agenda for development.

Despite the impressive global institutional efforts, high level meetings and agreements concerning sustainable development during the 1990s and early 2000s, very limited real progress has been achieved in terms of integration of the three main pillars of sustainability into the policy agenda, as confirmed during the debates and negotiations of the Agenda for Development post-2015 and the Sustainable Development Goals (SDGs), adopted in September 2015, and the crucial negotiations of the Paris Agreement on climate change.

The small island developing states (SIDS) have been considered a special case regarding the relation between environment (climate change) and development, owing to the particular challenges and options for this very heterogeneous group of countries.

The role of science has been crucial in assessing the need for an integrative approach for sustainable development and climate change responses, as revealed by the Fifth Assessment Report (AR5) of the Intergovernmental Panel on Climate Change (IPCC, 2014a). The next section of this chapter focuses on the relation between climate change responses and sustainable development, with reference to IPCC assessment reports (particularly the Fifth Assessment Report), including the IPCC review of socio-economic and environmental (climate change) scenarios. The section provides key inputs for exploring the challenges and options of the small island developing states in the subsequent two sections.

SUSTAINABLE DEVELOPMENT PERSPECTIVE IN THE IPCC FIFTH ASSESSMENT REPORT

The IPCC reports are integrated from the contributions of the three main working groups of this United Nations institution, created in 1988. Working Group (WG) I provides the scientific basis of climate change from the perspective of the natural sciences; Group II assesses the impacts, vulnerability and adaptation to climate change; and Group III deals with mitigation. The cycle of each evaluation report closes with a summary report, which includes basic results of the three groups. Following the practice of previous IPCC processes, this report has been a key reference document for multilateral political negotiations under the United Nations Framework Convention on Climate Change (UNFCCC) and the Kyoto Protocol.

The progress of the negotiation process for tackling climate change has been, however, extremely slow and fragmented, owing to the lack of political will of the major emitters to adopt equitable compromises, in correspondence with the historical and current levels of responsibility with regard to this global challenge.

IPCC findings have been very relevant for policy makers in small island developing states, particularly those conclusions relating to climate change implications for sea level rise and oceans, vulnerability of socio-economic sectors (tourism, agriculture, fishing, health, water resources and others) to climate change, and barriers and opportunities for improving adaptation capacity in a sustainable development context. However, in many cases, limited specialized literature relating to the SIDS restricts the scope and relevance of the corresponding assessments for these small territories.

Regarding the scientific basis of climate change, AR5 (IPCC, 2014a) states that the warming of the climate system is unequivocal, confirming what had already been concluded in AR4 (IPCC, 2007b). Furthermore, this new report highlights that atmosphere and ocean have warmed, the layers of snow and ice have diminished, sea level has risen, and concentrations of greenhouse gases (GHGs) have increased. Each of the last three decades has been successively warmer than previous decades, since 1850; and the period 1983–2012 was probably the warmest period of 30 years in the past 1400 years.

The results of WG I also note that human influence on the climate system is increasingly clear: that it is extremely likely (95–100%) that human influence has been the predominant cause of the warming observed since the mid-20th century.

This report pays special attention to the increasing concentrations of greenhouse gases since the Industrial Revolution in England (1750), which marked the beginning of the industrial age. Since that period, fossil fuels have had a dominant presence in the global energy sector. Today these fuels (oil, coal and natural gas) provide about 90% of the world commercial energy mix.

The report concludes that tackling climate change would require GHG emission reductions to be substantial and sustained. It is worth noting that most SIDS favor substantial global GHG emission reductions, owing to their high vulnerability to climate change.

Concerning the observed impacts and vulnerability, the IPCC WG II notes, among other findings, that changes in climate have caused impacts on human and natural systems on all continents and oceans in recent decades. It emphasizes that the differences in vulnerability and exposure are conditioned by non-climatic factors and multidimensional inequities that are often generated by processes of unequal development. These differences determine various derivative risks of climate change.

Chapter 29 of the WG II contribution to IPCC AR5 compiles relevant information on small islands. According to IPCC WG II, some low-lying developing countries and small island states are expected to face very high impacts that, in some cases, could have associated damage and adaptation costs of several percentage points of GDP.

Future risks of climate change will depend largely on how climate changes in the coming decades and centuries. Increasing magnitude of warming would most likely have serious and widespread consequences that may be unexpected and irreversible. Referring to future risks, the IPCC AR5 emphasizes those issues that represent major concerns for humanity, impacting in particular vulnerable regions and territories like the SIDS.

Adaptation as a key component of response strategies can contribute decisively to reduce the risks of climate change. On this subject, and in particular about management of future risks, the IPCC report notes that the adaptation is specific to each place and context, so there is no particular approach to risk reduction appropriate for all contexts. IPCC WG II highlights that adaptation to climate change generates a larger benefit to small islands when delivered in conjunction with other development activities, such as disaster risk reduction and community-based approaches to development.

According to the IPCC WG II, poor adaptation, which overstates the short-term results, or cannot sufficiently anticipate the consequences, could result in maladaptation; there is a gap between global adaptation needs and the funds available for this purpose; and there are co-benefits, significant synergies and trade-offs between mitigation and adaptation, and among adaptation responses, with interactions within and across regions. The report clarifies that climate resilient pathways are sustainable development pathways that combine adaptation and mitigation to reduce climate change and its impacts. IPCC WG II notes that adaptation and mitigation on small islands are not always trade-offs, but can be regarded as complementary components in responding to climate change.

Working Group III provides key findings regarding the conceptual framework for the mitigation of climate change and, in this regard, notes that effective mitigation would not be achieved if the individual agents put their own interests ahead. International cooperation is therefore required to effectively mitigate GHG emissions. The report reveals that the research and development (R&D) to support mitigation creates new knowledge, and international cooperation can play a constructive role in the development, diffusion and transfer of environmentally sound technologies and knowledge.

According to IPCC AR5, issues related to fairness and justice emerge with regard to mitigation and adaptation, so that results perceived as equitable lead to more effective cooperation. In addition, climate policy interacts with other social goals, creating the possibility of co-benefits or adverse side effects. IPCC WG II acknowledges that the ability of small islands to undertake adaptation and mitigation programs, and their effectiveness, can be substantially strengthened through appropriate assistance from the international community.

The new report highlights that global anthropogenic GHG emissions continued to grow during 1970–2010, with higher absolute growth per decade towards the end of this period. IPCC AR5 states that CO_2 emissions from combustion of fossil fuels and industrial processes contributed about 78% to the total increase in GHG emissions from 1970 to 2010, with a similar contribution in 2000–2010. Thus, CO_2 remains the main anthropogenic greenhouse gas, with 76% of the total in 2010, followed by methane (16%), nitrous oxide (6%) and fluorinated gases (2%). About half of the anthropogenic emissions of CO_2 accumulated between 1750 and 2010 occurred in the last 40 years.

According to IPCC WG III, in the absence of additional mitigation efforts beyond those existing today, emissions growth will continue to be driven by the dynamics in global population and economic activities. Reference scenarios, that is, those without additional mitigation efforts, show increases in global average surface temperature of between 3.7 °C and 4.8 °C by 2100, compared to pre-industrial levels.

The IPCC WG III pays special attention to emissions scenarios for this century. A set of 1200 scenarios was analyzed (including 900 mitigation scenarios), which provide important inputs about the economic, technological and institutional requirements of the different paths of mitigation and the implications.

According to the scenarios assessed, to limit temperature increase to 2 °C, a reduction of global greenhouse gas emissions by 40–70% in 2050, compared with 2010, and net negative or zero emissions of greenhouse gases by 2100 would be required, so that global emissions would peak within 5–15 years.

The IPCC WG III notes that only a small number of studies have explored scenarios in which it is likely that the change in temperature does not exceed 1.5 °C to 2100 relative to pre-industrial levels. These scenarios assume atmospheric concentrations below 430 ppm CO_2 equivalent, and therefore much more ambitious mitigation efforts. During the multilateral negotiations on climate change, including the negotiations of the Paris Agreement, SIDS strongly encouraged global efforts to keep average temperature increase below 1.5 °C, owing to their very high vulnerability to climate change even if the 2 °C goal is achieved.

This is a topic of key concern for the SIDS that needs further research and assessment work in the near future. For SIDS the 2 °C mitigation goal would imply serious socio-economic and environmental impacts, particularly for very vulnerable communities as well as fragile and unique ecosystems. Based on this consideration the Paris Accord of December 2015 invited the IPCC to provide a special report in 2018 on the impacts of global warming of 1.5 °C above pre-industrial levels and related global greenhouse gas emission pathways.

The IPCC WG III concludes that a delay of the additional mitigation efforts (beyond those existing today) by 2030 would substantially increase the difficulty of the transition to low emission paths in the long term and would narrow the range of options consistent with keeping the temperature change below 2 °C compared to pre-industrial levels.

The IPCC report reveals that there is a wide range of possible adverse side effects, as well as co-benefits and additional effects (spillover), derived from climate policies that have not been well quantified. Mitigation efforts and associated costs vary between countries and mitigation scenarios. Mitigation policies can devalue the assets of fossil fuels and reduce the income of fossil fuels exporters, but there are differences between regions and fuels.

The treatment of the nexus (between climate responses and sustainable development) has considerably improved since the IPCC Third Assessment Report (TAR) (IPCC, 2001). This process has been supported by an increasing participation in the IPCC working teams of authors and collaborators from the developing countries, as well as developed countries authors with a development theory perspective. IPCC AR5 shows the importance attached to the sustainable development perspective in the specialized literature assessed during that IPCC cycle, with 20 references to sustainability in the 31 pages of the IPCC AR5 Synthesis Report (IPCC, 2014a).

The recent discussion on climate change response options and sustainable development (SD) refers to several key aspects that have been widely examined by the scientific literature, including:

- Climate change as a challenge for SD.
- The integration of mitigation and adaptation into broader strategies for SD (realizing synergies and avoiding conflicts). This also considers the potential contribution of adaptation and mitigation to SD and equity, including poverty eradication.
- The role of SD in enhancing response capacities (for both adaptation and mitigation).

In this context, there is increasing recognition of the risks to small islands from climate-related processes originating well beyond the borders of an individual nation or island, as assessed by IPCC WG II.

According to IPCC AR5,

> adaptation and mitigation are complementary strategies for reducing and managing the risks of climate change. Substantial emissions reductions over the next few decades can reduce climate risks in the 21st century and beyond, increase prospects for effective adaptation, reduce the costs and challenges of mitigation in the longer term and contribute to climate-resilient pathways for sustainable development. (IPCC, 2014a: 17)

The most recent IPCC assessment (AR5) also concluded that

> many of those most vulnerable to climate change have contributed and contribute little to GHG emissions. Delaying mitigation shifts burdens from the present to the future, and insufficient adaptation responses to emerging impacts are already eroding the basis for sustainable development. Comprehensive strategies in response to climate change that are consistent with sustainable development take into account the co-benefits, adverse side effects and risks that may arise from both adaptation and mitigation options. (IPCC, 2014a: 17)

The above-mentioned conclusions regarding the relation between climate change and sustainable development are valid, in general, for all country groupings and regions, including the SIDS. IPCC AR5 reconfirms the high level of vulnerability of small islands to multiple stressors, both climate and non-climate.

The interaction between climate change response options and sustainable development has also been captured in the scenarios building process associated with the work of the IPCC, and the debate on response capacities to climate change, including adaptive and mitigative capacity, as part of a wider discussion on means of implementation of and barriers to sustainable development (IPCC, 2000, 2007a, 2007b, 2014a).

Both adaptive and mitigative capacities are closely interrelated as key components of the sustainable development capacity, which is in turn directly associated with human development, measured by the UNDP with three main dimensions (income, education and health). Thus, countries and regions with a higher Human Development Index (HDI) often have higher capacities for adapting to climate change and to mitigate it, while those areas with lower HDI have serious limitations in responding to climate challenges.

However, human development is a necessary but not sufficient condition for successfully responding to climate and other environmental challenges. Other factors not captured by the HDI, such as political willingness, effective institutional networks, citizen participation, and equity, play a decisive role in contributing to the capacities for responding to climate change. For instance, countries with a high HDI may find serious constraints in responding to extreme events in a context of slow institutional reactions and/or equity gaps, as was the case with experience of Hurricane Katrina in New Orleans, USA in 2005.

This section has provided key concepts and conclusions on climate change responses and sustainable development within the IPCC framework, including reference to key messages on: the physical basis of climate change; impacts, vulnerability and adaptation; and mitigation of climate change; as well as the notion of socio-economic and environmental (emissions) scenarios and the response capacity (adaptation and mitigation) of the various regions. The assessment of all these topics reveals general aspects of the special position of small island developing states, and this is further developed in the following sections.

CHALLENGES FOR SMALL ISLAND DEVELOPING STATES

According to the United Nations, SIDS include around 38 UN member small islands and low-lying countries,[1] which have a varied geography, climate, culture and level of economic development. They are scattered on the planet, but share priorities in terms of their economic, social and environmental development. As an expression of the interest surrounding the international discussions and negotiations on the status of SIDS, 2014 was declared International Year of Small Island Developing States.

SIDS were recognized for the first time, as a special case of the link between environment and development, at the United Nations Conference on this subject (Earth Summit) in Rio de Janeiro, 1992. Agenda 21, adopted at that conference, stresses that these states concentrate in a very limited space all the problems and challenges of the coastal areas. In fact, about 26% of the territory of SIDS is 5 meters or less above sea level; and about 30% of its 65 million people live in those areas (UNDESA, 2014b).

The institutional debate regarding SIDS has evolved considerably since the Earth Summit in 1992 and had a turning point with the adoption of the Barbados Plan of Action in 1994. Then the Mauritius Strategy of Implementation was agreed in 2005. These plans and strategies summarize the results of numerous meetings and discussions about the vulnerabilities, concerns and priorities of SIDS, and show the high degree of consensus about the requirements of sustainable development in those territories.

The Third International Conference on SIDS, which was held in September 2014 in Apia, Samoa, paid special attention to the challenges of SIDS in the current context of global economic crisis and increasing global environmental challenges. During the preparatory process for the conference in Samoa, the academic and political debate about the "blue economy" intensified. This concept tends to enhance the opportunities and manage the limitations of small island developing countries.

The concept of "blue economy" explicitly calls for a sustainable framework for developing countries, with emphasis on equitable access, development and distribution of the benefits of the resources of the seas and oceans, and the re-investment of these benefits in human development and debt relief. It also highlights key issues such as energy ("blue energy") and food security, among others. Moreover, the idea of international cooperation, based on science and research as a pillar of the "blue economy," and the principle of common but differentiated responsibilities are strengthened.

However, the interpretation of "blue economy" as an aspect of the "green economy" is polemic, as far as the first subscribes to the general principles of the second, that was subjected to heavy criticism before, during and after the Rio + 20 Conference in 2012.

Regarding the economic substance of this new concept, it is worth recalling that those options that may encourage extreme *commodification* of ocean resources (energy, minerals, etc.) must be avoided, as well as the risk of remaining trapped in conventional models, where the transfer of knowledge and technologies North–South is extremely limited. Similarly, with regard to so-called "blue carbon," significant risks may result from large-scale application of risky technologies such as geo-engineering for "carbon removal" of the oceans.

Economic Vulnerability

The priority areas of these countries in terms of economic activity include the dynamics and sources of growth of their economies, the behavior of their main trading and financial markets, and options for integration into the global economy.

Usually these territories, with small populations and very restricted domestic markets, have excessive dependence on international trade and finance, with significant barriers to taking advantage of economies of scale. Out of the 36 UN member SIDS that provide comparable statistics to the United Nations (2014), 20 of them have populations of less than 500 000 inhabitants, with lower levels in Pacific island states such as Tuvalu (10 000), Palau (21 000) and Marshall Islands (53 000), as well as St. Kitts and Nevis (55 000) in the Caribbean.

In several cases, low economic resilience is reflected in a high dependence on tourism and plantation agriculture, or external flows such as official development assistance (ODA) and remittances, as a source of employment and foreign exchange earnings.

Among those UN member SIDS that show higher levels of dependence on ODA as a percentage of GDP are Tuvalu (76.7%), Solomon Islands (49.6%), Sao Tome and Principe (30.2%), Haiti (23.2%), Tonga (21.3%), Samoa (16.4%), Cape Verde (13.6%), Guinea-Bissau[2] (12.3%) and Vanuatu (11.9%); and among the most dependent on remittances are Samoa (23.2% of GDP), Haiti (20.6%), Guyana[3] (16.5%), Jamaica (14.5%), Tonga (12.6%), Cape Verde (9.1%) and Timor-Leste (8.8%) (United Nations, 2014).

The economic vulnerability of these states also reveals the degree of isolation of their territories, that is, geographic dispersion and/or location in remote areas, which increases freight costs and reduces competitiveness. According to conservative estimates, the average remoteness of UN member SIDS from major international markets exceeds the global average by 18%, but this proportion exceeds 30% in several of these countries such as Tonga (58.9%), Fiji (56%), Vanuatu (54.5%), Samoa (51.7%), Solomon Islands (46.3%) and Mauritius (31.5%) (UNDP, 2014).

In addition, the island states show a high energy and food vulnerability, owing to the high dependence on imports of food and energy carriers (especially oil). These countries have experienced a rapid increase in energy consumption in recent decades as a result of industrialization, population growth and the expansion of rural electrification. They intensively use fossil fuels for power generation, industry and services.

SIDS are also economically affected by the high cost of infrastructure and public administration, including transport and communications, and the limited institutional capacity.

Social Challenges

Key issues for SIDS in the social sphere include their priorities in health, water and sanitation, and food security, as well as the situation of young people and women.

According to UNDP (2014), 20 out of 34 UN member SIDS considered are classified as countries with very high or high human development, nine have a medium level, and five a low level (Papua New Guinea, Solomon Islands, Comoros, Haiti and Guinea-Bissau). Thus, the economic fragility of the poorest nations in this group translates into significant challenges in the social sphere.

Life expectancy at birth is one of the key indicators in the calculation of the Human Development Index published by the UNDP (2014). On average, this indicator in the UN member SIDS is similar to the global average (71 years). From a group of 31 SIDS analyzed, 19 exceed the world average and 12 are below that level. The countries that show the best records of this indicator are Singapore[4] (82.2 years), Cuba (79.2), Maldives (77.7), Antigua and Barbuda (75.9) and Barbados (75.3); and the worst positioned are Papua New Guinea (62.3), Comoros (60.8) and Guinea-Bissau (54.2).

The most vulnerable segments of the population of SIDS are still women, children and indigenous communities, who are often the most affected by the low qualified health staffing that characterizes many of these countries. Twenty-five out of 27 UN member SIDS analyzed (UNDP, 2014) have fewer than 20 doctors per 10 000 inhabitants, and in 14 of them the proportion is less than 5 per 10 000. Cuba (with more than 67 doctors per 10 000 inhabitants) and Bahamas (28 per 10 000) are the SIDS with better provision of highly qualified health professionals. Timor-Leste (1 per 10 000), Guinea-Bissau (0.7) and Papua New Guinea (0.5) are the most critical cases in this field.

The coverage of water and sanitation is a fundamental dimension of human development, given its impact on the health sector. Most of the UN member SIDS have a potable water provision equal to or greater than 90% of their population (in 28 out of 36 countries analyzed for 2007–2011). The most favorable situation (100% coverage) corresponds to countries like Singapore, Barbados and Mauritius, and the most difficult conditions to nations such as Kiribati (66%), Haiti (64%) and Papua New Guinea (40%) (United Nations, 2014).

The provision of sanitation generally shows less favorable levels than the access to drinking water in these countries, with 14 out of the 36 UN member SIDS analyzed showing coverage below 80% of the population in 2007–2011. Singapore (100%), Palau (100%) and the Maldives (98%) are in the best positions, while Haiti (26%), Guinea-Bissau (19%) and Papua New Guinea (19%) present the most vulnerable situation.

In education, only eight out of the 21 UN member SIDS that offer statistics for 2005–2012 show literacy rates (adults) at or above the world average of countries with high human development, led by Cuba (99.8%), Antigua and Barbuda (99%) and Tonga (99%). Five countries in this group exhibited rates below 70%, with lower levels in Timor-Leste (58.3%), Guinea-Bissau (55.3%) and Haiti (48.7%) (UNDP, 2014).

Amid the current global trends, a group of these countries faces serious challenges for the replacement of the workforce required to boost their economies, as a combined result of trends such as the aging population and migration flows. Regarding the aging population, in 15 out of the 36 UN member SIDS analyzed, the proportion of people aged 60 and over within the total population is equal to or greater than 10%, especially in countries like Cuba (19%), Barbados (16.5%) and Singapore (16.4%). However, this ratio is less than 6% for countries like Belize,[5] Guinea-Bissau, Guyana, Timor-Leste, Solomon Islands, Papua New Guinea, Sao Tome and Principe, and Comoros (United Nations, 2014).

Environmental Challenges

The major environmental concerns for this group of countries are: climate change issues; the state of the oceans, seas and biodiversity; waste management; and energy challenges.

Multiple are the symptoms of environmental vulnerability affecting SIDS, including: their location in areas with great intensity and frequency of natural disasters such as hurricanes, earthquakes and volcanic eruptions; limited availability of natural resources and overexploitation and depletion of these resources (e.g. freshwater resources, soils); and high population density in coastal areas.

Some specialized sources (see UNDESA, 2014a) indicate that instead of SIDS they could be called "large oceanic states" because their territories represent a very small part of the total area under their jurisdiction. Under these circumstances, these states need to have sufficient capacity in terms of human resources and institutional networks for the sustainable management of the oceans, which constitutes a major challenge for them.

The oceans and seas, on which most of the population of SIDS basically depends, are affected by: marine pollution; the unsustainable exploitation of marine resources, including practices of illegal fishing and environmental erosion; physical alteration and destruction of marine habitat; and unsustainable coastal development.

Additionally, climate change tends to accelerate ocean acidification, coral bleaching, more intense extreme events, ocean warming and sea level rise, affecting key socio-economic sectors (e.g. agriculture, fishing, tourism, health) and compromising the very existence of some of these nations. Concerning the implications of sea level rise, IPCC AR5 WG II recalls that this is particularly important in small islands where the majority of human communities and infrastructure is located in coastal zones with limited on-island relocation opportunities, especially on atoll islands (IPCC, 2014b).

It is worth noting that the special conditions of small islands, regarding impacts, vulnerability and adaptation to climate change, have been assessed by the IPCC in the various assessment reports since the First Assessment Report (FAR) in 1990, with qualitative improvements across those reports. Specific chapters on small islands were included since the Second Assessment Report in 1995, with an increasing consideration of the interlinkages between climate change and development issues in subsequent assessments.

SIDS represent approximately 3% of the land area of the planet and provide habitat for about 20% of the plant species, reptiles and birds on the planet (UNDESA, 2014a). The loss of biological diversity affects economic resilience and food security, with serious implications especially for agriculture and rural communities. Bird species, for example, are facing the highest rate of extinction in SIDS as a result of anthropogenic influences (habitat destruction, overfishing and the introduction of invasive species).

The availability of fresh water is a serious concern in many of the island nations. In some SIDS, annual freshwater availability per capita is well below the 1000 cubic meters which is associated with a situation of water scarcity. Many of these nations are highly dependent on rainfall to feed surface intakes and replenish groundwater flows. Some SIDS, however, show significant freshwater resources, as in the case of Guyana (304.723 m^3 per capita), Suriname[6] (166.113 m^3), Papua New Guinea (114.217 m^3), Solomon Islands (83.086 m^3), Belize (50.588 m^3) and Fiji (32.895 m^3) (United Nations, 2014).

Freshwater resources are particularly sensitive to changes in climate. Changes in rainfall patterns can cause more severe and prolonged droughts, or may increase flooding. In addition, rising sea levels may raise the saltwater intrusion in coastal aquifers and affect the water quality of shallow wells.

For all the above, it is not by chance that SIDS have a very active agenda in multilateral negotiations on climate change. Many of these countries are members of the Alliance of Small Island Developing States (AOSIS), which calls for effective adaptation and far-reaching agreements for climate change mitigation. South–South cooperation may also contribute to enhancing the capacity of SIDS to respond to environmental (including climate change) and socio-economic challenges.

OPTIONS FOR SMALL ISLAND DEVELOPING STATES

The main opportunities and options for the small island developing states in the context of sustainable development are related to a great extent to the benefits of the sea and ocean resources, including services from marine and ocean ecosystems. It is worth recalling that oceans cover three-quarters of the planet's surface, and contain 97% of the water on Earth. Millions of species are in the oceans, but only around 200 000 of them have been identified.

Oceans also provide transport routes for 80% of the world trade volume, key resources for the tourism sector, 32% of the global supply of hydrocarbons, a vast potential of renewable energy (e.g. offshore wind, bioenergy), and abundant mineral reserves. The global market value of resources and industries related to marine and coastal activity is estimated at 3 trillion US dollars a year, representing about 5% of world GDP, and tourism generates about 1.3 trillion US dollars a year of income. In addition, marine biotechnology products have a market of 2.8 billion US dollars (2015 data), which could reach 4.6 billion US dollars in 2017. Marine ecosystems also provide more than 660 million jobs globally and are an important source of protein, especially for developing countries, with fishing contributing 15% of animal protein in our diets (FAO, 2015; United Nations, 2015a, 2015b).

Small island developing states benefit from the above-mentioned opportunities, but those economic advantages tend to decline as sea and ocean sustainability is adversely impacted by growing pollution, overexploitation of natural resources and climate change. In this context, a sustainable framework for these countries should consider, among other key goals: the equitable access, development and distribution of the benefits of sea and ocean resources; the re-investment of these benefits in human development and debt relief; international cooperation, based on science and research; and energy, food and water security.

Several examples of adaptation–mitigation interlinkages in small islands were assessed by the IPCC WG II (IPCC, 2014b), including energy supply and use, tourism infrastructure and activities, and functions and services associated with coastal wetlands.

According to the IPCC (2014b), the climate change response capacity of small islands can be substantially strengthened through appropriate assistance from the international community. IPCC WG II AR5 acknowledges that successful responses to climate change in these territories are also associated with empowering communities, optimizing the benefits of local practices, and working synergistically to progress development agendas.

The Paris Agreement of 2015 explicitly recognizes the special situation of the SIDS, jointly with the least developed countries (LDC), regarding the responses to climate change. This acknowledgement includes nine specific references to SIDS and LDC both in the Paris Agreement and in the corresponding Decision; and this is expected to facilitate the implementation of adaptation and mitigation actions in those groups of countries (see for reference Articles 4, 9, 11 and 13 of the Paris Agreement in UNFCCC, 2015).

Article 2 of the Paris Agreement refers to the Agreement objective, aiming to "strengthen the global response to the threat of climate change, in the context of sustainable development and efforts to eradicate poverty"; and Article 2a includes "holding the increase in the global average temperature to well below 2 °C above pre-industrial levels and to pursue efforts to limit the temperature increase to 1.5 °C above pre-industrial levels, recognizing that this would significantly reduce the risks and impacts of climate change."

The reference to the 1.5 °C goal in the text of the Paris Agreement is consistent with the priorities and needs of the SIDS.

CONCLUDING REMARKS

The small island developing states have been considered a special case, regarding the relation between the environment (climate change) and development, owing to the challenges and options for this very heterogeneous group of countries.

IPCC findings regarding the interlinkages between climate change and sustainable development are valid, in general, for all country groupings and regions, including the SIDS. IPCC AR5 reconfirms the high level of vulnerability of small islands to multiple stressors, both climate and non-climate.

Relevant IPCC findings on the SIDS include references to: climate change implications for sea level rise and oceans; vulnerability of socioeconomic sectors to climate change; and barriers and opportunities for improving adaptation capacity in a sustainable development context. However, in many cases, limited specialized literature on SIDS restricts the scope and relevance of the corresponding assessments for these small territories.

Challenges and opportunities for SIDS regarding the interlinkages between climate change and sustainable development have also been widely discussed in several international targeted conferences, including the Third International Conference on SIDS (in Apia, Samoa in 2014), as well as other broad United Nations processes such as the one leading to the adoption of the Agenda for Development post-2015 and the Sustainable Development Goals (SDGs).

Prioritized areas for these countries, in terms of economic activity, include: improving the dynamics and enhancing the sources of their economic growth; increasing the knowledge on their main trading and financial markets; and exploring new options for integration into the global economy. Key issues for SIDS in the social sphere include health, water and sanitation, food security, and the situation of young people and women. The major environmental concerns for this group of countries are: climate change issues; the state of the oceans, seas and biodiversity; waste management; and energy challenges.

The special conditions of small islands regarding impacts, vulnerability and adaptation to climate change have been assessed by the IPCC in the various assessment reports since 1990, with qualitative improvements across those reports. Specific chapters on small islands have been included since the Second Assessment Report in 1995, with more comprehensive reviews of the interlinkages between climate change and development issues in subsequent assessments.

The main opportunities and options for the SIDS in the context of sustainable development are related to the benefits derived from the sea and ocean resources, including services from marine and ocean ecosystems. Some specialized UN sources indicate that instead of SIDS they could be called "large oceanic states," because their territories represent a very small part of the total area under their jurisdiction.

A sustainable framework for SIDS, should consider, among other key goals: the equitable access to, development of and distribution of the benefits of sea and ocean resources; the re-investment of these benefits in human development and debt relief; international cooperation, based on science and research; and energy, food and water security.

According to Chapter 29 of the WG II contribution to IPCC AR5, adaptation to climate change generates larger benefit to small islands when delivered in conjunction with other development activities, such as disaster risk reduction and community-based approaches to development.

The Paris Agreement of 2015 also recognizes explicitly the special situation of the SIDS, regarding the responses to climate change. This acknowledgement includes nine specific references to SIDS in the Paris Agreement and the corresponding Decision; and this is expected to facilitate the implementation of adaptation and mitigation actions in SIDS.

Areas deserving further consideration to improve policy-relevant scientific research on SIDS challenges and options, regarding the environment (climate change) and sustainable development, include – inter alia – the following requirements:

- increasing regional, sub-regional and thematic scientific research with a SIDS perspective, including intra-SIDS joint research and modelling efforts;
- improving the participation of SIDS experts on international research or policy-relevant assessment exercises (e.g. IPCC, IPBES);
- enhancing the communication of scientific research and policy-relevant pilot experiences among SIDS;
- receiving additional international support to increase SIDS research capacities;
- fostering the role of local science in informing relevant policies in this field.

NOTES

1. Antigua and Barbuda, Bahamas, Bahrain, Barbados, Belize, Cabo Verde, Comoros, Cuba, Dominica, Dominican Republic, Fiji, Grenada, Guinea-Bissau, Guyana, Haiti, Jamaica, Kiribati, Maldives, Marshall Islands, Mauritius, Federal States of Micronesia, Nauru, Palau, Papua New Guinea, St. Kitts and Nevis, St. Lucia, St. Vincent and the Grenadines, Samoa, Sao Tome and Principe, Seychelles, Singapore, Solomon Islands, Suriname, Timor-Leste, Tonga, Trinidad and Tobago, Tuvalu, Vanuatu. The UN information also mentions 20 non-UN members/associate members of the Regional Commissions: American Samoa, Anguilla, Aruba, Bermuda, British Virgin Islands, Cayman Islands, Commonwealth of Northern Marianas, Cook Islands, Curacao, French Polynesia, Guadeloupe, Guam, Martinique, Montserrat, New Caledonia, Niue, Puerto Rico, St. Maarten, Turks and Caicos Islands, US Virgin Islands (see: http://unohrlls.org/about-sids/country-profiles/).
2. Guinea-Bissau is a small low-lying state on the African mainland, but owing to its vulnerability to sea level rise it is officially included in the group of SIDS, according to the United Nations.
3. Guyana is a small low-lying state on the South American mainland, but owing to its vulnerability to sea level rise it is officially included in the group of SIDS, according to the United Nations.
4. Singapore is a small high-income island state officially included in the group of SIDS, according to the United Nations.
5. Belize is a small low-lying state on the Central American mainland, but owing to its vulnerability to sea level rise it is officially included in the group of SIDS, according to the United Nations.
6. Suriname is a low-lying state on the South American mainland, but owing to its vulnerability to sea level rise it is officially included in the group of SIDS, according to the United Nations.

REFERENCES

FAO (2015), '100 facts in 14 themes linking people, food and the planet', http://www.fao.org/fileadmin/user_upload/mdg/100_facts/100facts_EN.pdf.

IPCC (2000), *Special Report Emissions Scenarios*, Special Report of IPCC Working Group III, Summary for policymakers, www.ipcc.ch.

IPCC (2001), *Climate Change 2001: Synthesis Report*, Summary for policymakers, approved at IPCC Plenary XVIII, Wembley, UK, 24–29 September, www.ipcc.ch.

IPCC (2007a), *Towards New Scenarios for Analysis of Emissions, Climate Change, Impacts, and Response Strategies*, IPCC expert meeting report, Noordwijkerhout, the Netherlands, 19–21 September, www.ipcc.ch.

IPCC (2007b), *Climate Change 2007: Synthesis Report*, Summary for policymakers, approved at IPCC Plenary XXVII, Valencia, Spain, 12–17 November, www.ipcc.ch.

IPCC (2014a), *Climate Change 2014: Synthesis Report*, Summary for policymakers, www.ipcc.ch.

IPCC (2014b), *Climate Change 2014: Impacts, Adaptation, and Vulnerability*, IPCC Working Group II contribution to AR5, www.ipcc.ch.

UNDESA (2014a), *Partnerships Briefs for Small Island Developing States: Oceans, Seas and Biodiversity*, UN Conference on Small Island Developing States, Division for Sustainable Development.

UNDESA (2014b), *Partnerships Briefs for Small Island Developing States: Climate Change and Disaster Risk Management*, UN Conference on Small Island Developing States, Division for Sustainable Development.

UNDP (2014), *Human Development Report 2014*, New York: UNDP.

UNFCCC (United Nations Framework Convention on Climate Change) (2015), *Adoption of the Paris Agreement*, Conference of the Parties, Twenty-First Session, Paris, 30 November – 11 December, FCCC/CP/2015/L.9/Rev.1, 12 December.

United Nations (2014), *Population and Development in SIDS*, Department of Economic and Social Affairs, Population Division, www.unpopulation.org.

United Nations (2015a), *Blue Economy Concept Paper*, https://sustainabledevelopment.un.org/content/documents/2978BEconcept.pdf.

United Nations (2015b), 'Sustainable Development Goals: 17 goals to transform our world. Goal 14: Conserve and sustainably use the oceans, seas and marine resources', http://www.un.org/sustainabledevelopment/oceans/.

PART II

Climate resilience: sectoral perspectives

7. Can agriculture be climate smart?

Irina Arakelyan, Anita Wreford and Dominic Moran

INTRODUCTION

Climate change is already having a significant impact on agriculture through greater weather variability and the increasing frequency of extreme events, both of which are expected to increase over the coming decades (FAO, 2013a). As a largely human-made adjunct to the natural environment, the sector is highly exposed to these changes, and much of the international climate discourse has focused on the sector's vulnerability, resilience and adaptive capacity for given projected climate scenarios (Nelson et al., 2009). As in other sectors the agricultural adaptation agenda has highlighted the uncertainty of impacts, the role of research and development, and the likelihood that there will be winners and losers. It has also emphasized the specific vulnerabilities of low income smallholders in developing countries and ethical reasons why the developed world should invest more to help low income countries adapt.

Agriculture also has a recognized role in terms of climate change mitigation. The sector accounts for approximately a third of total global anthropogenic greenhouse gas emissions (including emissions from land use change and deforestation) (IPCC, 2014). Farmers and land managers have a significant role to play because emissions reduction measures can be taken to increase soil carbon sequestration, manage fertilizer application, and improve ruminant nutrition and waste. The global significance of such actions should not be underestimated. Existing research (Moran et al., 2011) shows that some of these measures are low cost relative to the costs of reducing emissions in other sectors such as energy or heavy industry. Indeed, avoiding excessive fertilizer application alone could abate 2–6 per cent of China's total emissions of nitrous oxide, a powerful greenhouse gas, while saving significant input costs borne by farmers (Wang et al., 2014). But the mitigation potential is also hindered by institutional and behavioural barriers retarding the adoption of these measures in developed and developing countries. This includes formal agreement on how agricultural mitigation should be treated in national obligations, commitments or targets.

These climate challenges overlap a growing concern about global food security, which highlights additional stressors including demographic changes, natural resource scarcity, and economic convergence in consumption preferences, particularly livestock products. The Food and Agriculture Organization (FAO) of the United Nations estimates that, to meet the demands of a larger population (expected to reach 9 billion by 2050), food supply will need to grow by 60 per cent (FAO, 2012). Other estimates suggest an increase of 70–100 per cent (e.g. Godfray et al., 2010). This imperative has led to calls for a new green revolution, or a contemporary equivalent, sustainable intensification (SI), which implies a step-change to do more with less while minimizing harmful externalities. The concept of climate smart agriculture (CSA) is arguably nested within the SI concept, predominantly addressing the climate dimensions and imperatives and reconciling these with productivity objectives, predominantly in developing countries. How these objectives can be balanced and what they imply for farming systems in developing and developed country farm systems have recently become the focus of much scrutiny.

This chapter explores the definition of CSA, and what the concept offers to pre-existing sustainability debates. We consider some emerging contradictions and opportunities and some of the political economy challenges in using CSA as the route to resilience in global agriculture. The chapter goes on to outline some of the definitions and critiques of CSA, before moving on to an overview of CSA measures. It provides some examples of the current application of climate smart interventions, before discussing policies and institutions for achieving climate smart agriculture. Conclusions and implications are then presented.

BACKGROUND AND BROAD DEFINITIONS OF CSA

The CSA term was first coined by FAO as 'agriculture that sustainably increases productivity, resilience (adaptation), reduces/removes GHGs (mitigation), and enhances achievement of national food security and development goals' (Lipper et al., 2010, p. ii). The concept nominally addresses the challenges of climate change in agriculture while simultaneously trying to increase food security. CSA therefore encompasses a variety of farm practices including soil management, crop selection, livestock management and energy use measures (Table 7.1). It is important to note that there is no definitive taxonomy of CSA measures for the food supply chain (i.e. within and beyond the farm gate). Instead, CSA interventions

Table 7.1 Examples of potential CSA measures

Measure	Adaptation/resilience	Mitigation	Productivity
On-farm tree planting	Shelter/shade for livestock.	Carbon sequestration.	Reduced heat stress for livestock.
Planting hedgerows and buffers	Preventing drought through reducing runoff.	Carbon sequestration.	Avoided pasture/crop loss.
Soil management practices	Increased crop/pasture resilience.	Increased soil organic carbon.	Avoided pasture/crop losses.
Managing animal health and disease	Avoid disease outbreaks.	Unwell animals are less efficient and emit more methane per unit.	Avoided mortality and illness.
Livestock diet management	Appropriate diet can reduce heat stress.	Appropriate diet can reduce methane emissions.	Avoided mortality or reduced production.
Natural flood management (e.g. woodland and peatland restoration, riparian planting)	Reduced flood damage.	Carbon sequestration.	Avoided or reduced pasture/crop/livestock loss.

will be as varied as the diversity of farm regions, with applicability varying according to the type, location and scale of the production system. To some extent many of these measures have long been part of traditional farm practices in many parts of the world (Richards et al., 2014). In terms of implementation, measures should be judged by clear criteria of technical effectiveness, cost-effectiveness (or economic efficiency) and social acceptability. Note also that, while there is a general view of these measures applying within the farm gate, the use of some (e.g. reduced use of synthetic fertilizer) will imply wider supply chain effects that need to be incorporated into any evaluation of CSA costs and benefits.

Steenwerth et al. (2014) provide an extended definition of CSA, including provision for food security, poverty reduction and contributing to economic development. The authors also mention increasing the productivity and resilience of agricultural ecosystem functions, as well as negotiating trade-offs in meeting these objectives. In this definition CSA is described as a continuous process of adaptive management in which stakeholders, researchers and policymakers come together to meet the challenges of climate change and help transform agricultural and food systems towards sustainability goals. This is obviously a more aspirational long-term agenda, which is less specific about what measures to apply where, and what transformation actually means in developed and developing country food systems. Nevertheless a major focus is on smallholder farmers in developing countries as a group in most need to increase food security.

CSA has gained significant traction among governments of developing countries, as well as bilateral and international donor agencies. In the latter case the concept has fitted seamlessly into the overlapping space between aid for rural poverty, pro-poor climate (adaptation) assistance and sustainable development more generally. CSA projects around the world have been facilitated by the FAO and the World Bank and their partners. The importance of the approach on the global agricultural development agenda has been endorsed with the foundation of the Global Alliance for Climate-Smart Agriculture launched in September 2014 at the United Nations climate summit. The Alliance was set up as a 'voluntary, farmer-led, multi-stakeholder, action-oriented coalition committed to the incorporation of climate-smart approaches within food and agriculture systems' (Climate Summit, 2014).

Despite these developments, the CSA concept has come under more critical scrutiny from major non-governmental organizations (NGOs), charities, some farmer organizations and governments. Criticism has focused on the potentially disproportionate emphasis given to mitigation in CSA, as well as the focus on growth, and the long-term sustainability of the entire approach. There is also concern regarding the perceived misappropriation of the term by multinational agribusinesses seeking to rebrand existing practices under a more benign label taking advantage of the fluidity of definitions and lack of criteria (e.g. Civil Society, 2015), where CSA can mean everything and nothing simultaneously (Lilliston, 2015). Some opponents of CSA appear to prefer the concept of 'agro-ecology', which has an established scientific grounding, and provides a response not only to climate change but also to wider environmental and social challenges facing agriculture. When it comes to smallholder farming systems, there is also a more mundane debate about whether CSA measures can or will, as assumed by some aid donors, actually be adopted and whether specific information and incentives need to be improved to increase uptake. This is reminiscent of an older debate on technology adoption in global agriculture, drawing attention to the need to consider the nature of supply chains beyond the farm gate and how these can influence the on-farm choices made by farmers.

CSA CRITIQUE

The appeal of CSA as a driving concept by the range of organizations and governments described previously is largely due to the inclusivity of the term and its broad aims. The promotion of productivity as a pillar alongside mitigation and adaptation accords with a productivist market-oriented focus for the sector that is common in most developed countries. An acknowledgement that this agenda can easily accommodate climate change obligations and challenges without radical policy changes means that the term is politically palatable, or at least unthreatening, to the agricultural lobby. At the same time, the vagueness of the term allows a broad range of stakeholders to take ownership, at least nominally.

The main objectives of CSA throw up individual critiques that warrant scrutiny for their practical, political and ethical framing in both developed and developing countries, and more generally because agriculture is often seen as exceptional when it comes to addressing climate change. Fundamentally agriculture is exceptional in the sense that it comprises thousands of small businesses each working in biophysically different and often challenging conditions, with individual farmer behaviours driven by a diversity of economic and social motives. These conditions mean that there is considerably greater uncertainty about the technical effectiveness of either mitigation or adaptation measures when applied in some farming conditions. This also implies that governments tend to avoid regulation that can be costly to implement and monitor for an uncertain outcome.

Exceptionalism has also been used to argue that GHG mitigation cannot be a priority for farmers in a world where food production arguably needs to increase. This argument would possibly be more robust were there a more even distribution of current production, or if there were less food waste, but the logic has nevertheless gained considerable traction among politicians in many countries with strong agricultural lobby groups. Much the same argument can be made for diverting existing support or development assistance towards agricultural adaptation, which can be branded as climate smart, while apparently targeting food and livelihood security. Both arguments are possibly worth challenging in the sense that most adaptations and some types of mitigation are often in the best self-interest of any individual producer. That is, adaptation measures aim to avoid losses resulting from climate impacts, and some mitigation measures are demonstrably win–win in terms of saving input costs and reducing emissions (or vulnerability). Perhaps the more subtle essence of CSA therefore is to identify specific current or future double- or triple-win measures and to tease out where climate and production objectives can be adopted as part of current or traditional 'good' farm practice. Put another way, the main emphasis of CSA is about overcoming a form of information failure that impedes the adoption of relevant measures. Resolving this failure is a shared objective of both public and private sector stakeholders.

In the context of emissions mitigation, the aforementioned arguments tend to be common in both developed and developing countries. In the former where agriculture is an important contributor to the economy (New Zealand and Ireland being specific examples), but a significant source of emissions, mitigation obligations tend to be resisted because of their perceived potential effect on production. Here the rhetoric tends to focus on increased efficiency and thereby reducing emissions per unit of product rather than attempting to reduce overall emissions from the sector (e.g. New Zealand Agricultural Greenhouse Gas Research Centre, 2014). While efficiency is undoubtedly laudable, some evidence suggests that any emissions savings from efficiency will likely be offset by increased production, the so-called Jevons' paradox (Alcott, 2005); and ultimately emissions need to fall in absolute terms if dangerous climate changes are to be prevented (IPCC, 2014). These countries argue further that, if they reduce their (efficient) production, other, less efficient producers will simply fill the supply gap, leading to an increase in global emissions, the carbon leakage argument (Lee et al., 2006). Meanwhile, as noted, low income countries, particularly those with subsistence farmers who struggle to meet their own food requirements, also argue that mitigation as a largely global problem is less urgent than national food security. In this regard there is concern among some organizations (e.g. see Civil Society statement signatories in Civil Society, 2015) that development aid is being repackaged and made conditional on projects being climate smart and meeting all three pillars of CSA, leading to increased vulnerability among smallholder farmers.

This lack of conceptual clarity extends to some of the defining practical attributes of CSA and the nature of policies to achieve balanced outcomes or to mediate trade-offs between the objectives. The issues are both spatial and temporal. The spatial challenge relates to whether the three objectives need to apply simultaneously at the farm scale or whether within-sector specialization can occur to allow some producers to focus on increasing mitigation potential and others to focus on intensive production. The temporal challenge relates to the timescale over which goals are considered, which may alter the outcome of the measure. For example, practices that increase productivity in the short term may be unsustainable in the longer term, possibly even leading to reversals in productivity, particularly if adaptation or resilience is not also considered. Alternatively, measures to increase soil carbon may reduce emissions in the short term but may be less stable over the longer term once carbon storage has reached a saturation point, usually within 20 to 100 years, after which net emissions from the sector will increase (Climate Focus, 2011).

A further institutional issue is how farmers are rewarded for above-baseline (or business as usual) activities such as carbon sequestration. On current evidence many farmers are resistant to climate change arguments, and there is limited evidence demonstrating how reducing GHGs can be synergistic with saving costs and adapting farm systems (double or triple wins). Accordingly there is a need for some incentive to adopt climate smart measures. Since most of agriculture is currently outside existing carbon markets, that could conceivably give farmers the incentive to measure their emissions and offer them for sale. Activities need to be developed in relation to the voluntary carbon market or in anticipation of public (i.e. government) payments for ecosystem services. In either case there is a need for baseline information and for the measuring, reporting and verification (MRV) of performance by institutions in terms of inputs or outcomes. Such institutions have yet to be developed in any country, although there is considerable ongoing research to develop simplified farm calculators for emissions.

CURRENT APPLICATION OF CSA MEASURES

All CSA measures face multiple challenges because of their inherent complexity (Shames et al., 2012). MRV has already been noted as a challenge when attempting to measure the mitigation benefit across a large population of smallholders. There is a general lack of consensus on measurements in climate smart agriculture, particularly when it comes to choosing quantitative indicators, as resource limitations prevent more thorough comparisons on the advantages of any particular system. The lack of an agreed MRV methodology as a common basis for the offset credits (Sharma and Suppan, 2011), and international understanding of MRV concepts and practical design, is limited (Climate Focus, 2011; Sharma and Suppan, 2011). All this leads to large uncertainties with quantification (Gattinger et al., 2011).

Several further concerns have been raised about the sustainability of climate smart investments by smallholders who might also have insecure land tenure. This often prevents them from obtaining the credit necessary to become involved in or to sustain a CSA activity and securing potential benefits from CSA projects in the long term (Hilger et al., 2013; McCarthy and Brubaker, 2014).

Despite numerous donor-funded climate smart projects being implemented globally, there is limited evaluation evidence to judge the outcomes (Westermann et al., 2015). Most of the available literature reports only anticipated outcomes rather than actual results, partly because many of the projects are still in the implementation stage. To date the majority of this evidence comes from projects implemented in sub-Saharan Africa.

Promoting Conservation Agriculture in Malawi and Zambia

Conservation agriculture (CA) is promoted as one of the most important, triple-win CSA practices, and includes the three core principles of: minimum soil disturbance; permanent soil cover; and crop rotation. CA has been shown to reduce soil erosion and retain water in the soil, hence leading to higher crop yields (McCarthy and Brubaker, 2014). However, emerging evidence shows that the ability of CA to achieve higher yields and sequester carbon is highly site-specific, dependent on the weather and fertilizer, as well as herbicide and labour availability (Giller et al., 2009; Thierfelder et al., 2012; Andersson and D'Souza, 2014). In Malawi, for example, nearly three-quarters of CA promotion projects promote the use of herbicides to save labour (through the increased need to do weeding), despite their high cost and limited availability (Andersson and D'Souza, 2014), which leads to questions about the role of partnering herbicide companies in these projects (Ito et al., 2007). Adoption of CA involves long-term investment costs including purchase of necessary tools (such as hand tools, animal traction or tractor equipment, particularly for direct seeding and planting in non-tilled soil), which are not readily available in many markets in sub-Saharan Africa (Meybeck and Gitz, 2012), as well as high opportunity costs due to diverting residues from other traditional uses (e.g. the use of crop residues as feed for animals). More importantly, while CA sometimes increases yields in the long term, farmers may need to wait three to seven years (some studies suggest up to ten years) to see yield increases (McCarthy and Brubaker, 2014; Richards et al., 2014). Overall, the evidence suggests that CA can often generate negative outcomes, including no yields or even reduced yields for the first five to ten years of practising (Gattinger et al., 2011; McCarthy and Brubaker, 2014), while the high costs of adoption affect projects from early on. In addition, benefits of adoption are highly dependent on agro-ecological conditions and sustained investments (Pervin et al., 2013).

Further, the role of CA in carbon sequestration has been questioned, as evidence is lacking or inconclusive (Andersson and D'Souza, 2014). Even if benefits do occur, they are restricted to the upper 10 cm of the soil, while no sequestration – or even a carbon deficit – has been found at soil depths below 20 cm (Gattinger et al., 2011).

In 2012, the FAO and European Commission (EC) introduced a €5.3 million three-year programme 'Climate Smart Agriculture: Capturing the Synergies among Mitigation, Adaptation and food Security'. The main aim of the programme was to strengthen the technical, policy and investment capacities of its three partner countries – Malawi, Vietnam and Zambia – to enable sustainable increases in agricultural productivity and incomes, resilience of agricultural and food systems to adapt to climate change, and opportunities to reduce and remove GHGs in order to meet their national food security and development goals (FAO, 2014). One of the flagship projects of this partnership was the promotion of conservation agriculture in Malawi and Zambia.

Initial project findings showed that many farmers in the two countries had difficulties adopting CA, with reasons including the need to use crop residues for animal feed instead of soil cover, and many of the farmers being too poor to wait for the promised increases in yields for at least several seasons (FAO, 2013b, 2014). Although the adoption rates of CA in Zambia are relatively higher than in other sub-Saharan African countries, these are still lower than expected, with many subsequently abandoning the measure (Kaczan et al., 2013). The findings showed only modest evidence of yield benefits, as well as financial and livelihood benefits of conservation agriculture, in Zambia, all of which are achievable in the medium to long term (Kaczan et al., 2013). The findings also showed considerable constraints to the adoption of conservation agriculture in both countries, with the most important being labour constraints, resulting in a very limited adoption of the practice.

Agricultural Carbon for Smallholder Farmers in Kenya

The Kenya Agriculture Carbon Project (KACP) is the flagship World Bank project on agricultural carbon finance in Africa, which was set up in 2010 to increase agricultural productivity and encourage private sector investment in agricultural enterprises but with the addition of a carbon component (World Bank, 2012–2014). Carbon sequestration activities include various elements of conservation agriculture including reduced tillage, cover crops, residue management, mulching, composting, green manure, targeted application of fertilizers, reduced biomass burning and agroforestry. The project aims to provide technical support to about 60000 farmers aggregated in farmer groups, managing a total of 45000 hectares in the Nyanza and Western provinces in Kenya. The project is implemented by the Swedish Cooperative Centre, ViAgroforestry, and uses an input-based carbon-accounting methodology.

According to the World Bank, improved agricultural practices for this project have the potential to sequester about 60000 tons of CO_2-equivalents per year, while simultaneously increasing crop yields, diversifying income sources and reducing the vulnerability of small farmers to climate change (World Bank, 2012–2014). The carbon credits generated by this project are to be purchased by the BioCarbon Fund, which is a public–private initiative administered by the World Bank.

However, the evidence shows that the targeted number of farmers has not been achieved, with the target subsequently being revised downwards (Lang, 2011). The total financing for the project amounted to US$2.5 million, 40 per cent of which will go to various intermediaries in transaction costs. Farmers involved in the project would only receive an average of around US$1 per year over 20 years. Further, the exact amount of carbon that will be sequestered is still uncertain (Lang, 2011). Although the World Bank reports stated that the main benefits will be increased yields (in the long term) and reduced vulnerability to climate change, rather than the carbon payments, the failure to get many farmers on board the project, and the increasing dropout rate, shows that the dominant donor-driven narrative about 'triple wins' does not always resonate well with local circumstances (Sharma and Suppan, 2011; Atela, 2012).

Natural Regeneration in Niger and Ethiopia

Agroforestry (AF) is another farming technique that is well aligned with the goals of CSA. Agroforestry refers to land use practices in which trees and shrubs are integrated with agricultural crops (Branca et al., 2011).

Multiple benefits of agroforestry have been documented, including increase and diversification of farm income, minimizing the effects of extreme climate events such as erratic and heavy rains and droughts, prevention of erosion, and improvement and enrichment of land (Meybeck and Gitz, 2012). Overall, a successful agroforestry system can help farmers adapt to climate change, improve their livelihoods by increasing yields and diversifying their incomes, and sequester significant amounts of carbon via extended tree cover (Mbow et al., 2014).

Eighty-four per cent of the population in Niger depends on agriculture for survival, and half the population has a low level of food security (Pye-Smith, 2011). In the 1990s, the government revised its forestry regulations, giving farmers the right to decide how to manage trees on their land. Farmers started planting *Faidherbia albida*, a tree which is known for improving soil fertility. Other species of trees were also planted, providing fruit, fodder for livestock and fuelwood. This tree expansion programme spread, resulting in a major transformation of landscapes, especially in the Maradi and Zinder regions of Niger. After more than two decades, over 5 million hectares of land have been transformed, benefiting 4.5 million people. In participating areas, sorghum yields have increased by 20–85 per cent and millet yields by 15–50 per cent. The trees diversify farmers' incomes and provide alternatives to famine in case of drought. The system is also sequestering significant quantities of carbon (although this has not been quantified), mainly owing to its aggregated impacts, and helps farmers adapt to climate change (Pye-Smith, 2011).

In Ethiopia, the over-exploitation of forest resources has affected more than 95 per cent of the country's native forests (Pye-Smith, 2011). Deforestation has caused severe erosion, resulting in floods and affecting groundwater reserves which provide people with potable water. In Humbo region in Ethiopia, where the majority of the population depend on agriculture for their livelihoods, an increased occurrence of droughts and floods both as a result of climate change and as a consequence of reduced tree cover can increase vulnerability. Humbo Assisted Natural Regeneration Project, facilitated by World Vision and the World Bank, established seven forest cooperatives to sustainably manage and reforest the surrounding land. Since the start of the project, more than 90 per cent of the Humbo project area has been reforested using the farmer-managed natural forest regeneration technique. This has resulted in increased production of wood and tree products, as well as fodder, and an increase in incomes. Furthermore, the regeneration of the native forest has led to reduced soil erosion and flooding, and provided an important habitat for the local biodiversity. Owing to the project's mitigation impact it has become the first large-scale forestry project in Africa to be registered with the United Nations Framework Convention on Climate Change (UNFCCC).

Although not originally intended as climate smart, the last two examples highlight that buy-in and ownership by local communities are essential for any climate smart project to be sustainable and successful in the long run. It is clear that economic, social, cultural and environmental trade-offs are unavoidable in order to achieve climate smart outcomes. But, following an integrated approach, some of these practices can increase long-term productivity, create synergies between adaptation and mitigation initiatives, and reduce the necessary trade-offs (Bromhead et al., 2013).

POLICIES AND INSTITUTIONS FOR ACHIEVING CLIMATE SMART AGRICULTURE

CSA cross-cuts the existing agricultural policy landscape that has historically focused on production in both developed and developing countries. Climate change is a relatively new consideration for agriculture, and many countries have assigned a low priority to defining its mitigation potential and to rolling out adaptation practices. This can be explained in part by the biological complexity previously mentioned and also lack of national and global institutional structures governing mitigation and adaptation policy. For example, agricultural mitigation is not central to UNFCCC submissions, and is generally kept outside of mandatory emissions reductions targets in most countries. Governments can choose to reward mitigation using existing support policies where these exist, but the uncertainties surrounding the quality and permanence of agricultural mitigation (plus MRV costs) mean that few countries are making much progress. The same problems are delaying the development of carbon credits in the sector and the penetration of carbon trading more generally in agriculture.

Thinking more broadly, vulnerability of global food systems is a real possibility, and this highlights a more general lack of global governance of global food systems. Accordingly, does the global food security need suggest that there should be a global architecture supporting CSA practices? What would this look like? Evidence from historical and existing policy interventions illustrates the difficulty of achieving policy coherence and avoiding unintended consequences. Well-meaning policies intended to support vulnerable farmers to maintain production under challenging circumstances, notably the use of input subsidies (for example fertilizer subsidies in Indonesia (Osorio et al., 2011) and water subsidies in Spain and India (GSI, 2010)), often result in continued production in areas that are no longer appropriate, creating barriers to adaptation and transformation, as well as a host of environmental problems, including increased emissions, and are generally known to be inefficient and inequitable mechanisms for transferring resources. Environmental-focused policies can also lead to serious distortions: biofuel policy for example, aiming to reduce emissions from fossil fuel combustion, has led to a number of 'emergent' risks, including increased food insecurity, through competition with food crops for land, as well as several environmental concerns (see Oppenheimer et al., 2014 for a summary). And policies that aim to increase resilience to climate changes may in the short term have negative effects on productivity as well as GHG emissions (for example extensification and/or diversification).

So would it be best to avoid wading in with enthusiastic policy instruments to facilitate CSA, and hope that market-based solutions will emerge? Certainly, the development of carbon prices and credits for mitigation measures would be a powerful incentive for the development of MRV protocols in developed and developing countries. Countries are unwilling to act first (New Zealand, for example, has stated that it will only introduce emissions reduction obligations if its international competitors also do so); therefore a level playing field – achieved through binding commitments, at least within developed countries, to introduce carbon charges on agriculture – is essential. And, before that, the UNFCCC method of recording emissions must be adjusted so that farmers are given credit for the changes they do make. But there is no guarantee that the resulting systems will be climate resilient.

There is general agreement that longer-term capacity building is more effective than a focus on short-term specific actions, so that local actors are able to make appropriate decisions for their individual circumstances. Rather than creating new policies, communicating the basic principles and aims of CSA and removing potential barriers, and then allowing actors at different scales to make their own decisions, may be the most effective approach. Adaptation already tends to be integrated or 'mainstreamed' into existing policies (e.g. as outlined in the Adaptation White Paper/EU Adaptation Policy), in a recognition of this principle.

CONCLUSIONS

CSA is the latest attempt to reconcile the dual, sometimes competing, prerogatives for achieving global food security: increasing agricultural productivity whilst maintaining environmental integrity through a combination of mitigation and adaptation strategies. The practical results of these dualistic trade-offs often create unforeseen consequences, such as an increased reliance on technological and other inputs. This can further marginalize those who are already excluded from the global development agenda, thereby exacerbating existing inequalities in the food system.

However, even bearing in mind that some emissions in agriculture are unavoidable, agriculture can be climate smart under certain conditions. To facilitate this, it will be important to identify specific current or future triple-win measures, and here we note the limited empirical evidence so far to support the concept of triple wins in climate smart agriculture.

Further, it will be essential to explore where climate and production objectives can be adopted as part of current or traditional 'good' farm practice and to shift the focus from a short-term increase in on-farm productivity to the longer-term and more sustainable approach of improving supply chain efficiency. It is also important to keep in mind that survival remains the main priority of millions of smallholder farmers in the developing world, and mitigation objectives of CSA could potentially be more achievable (and yield more results) in large-scale commercial agriculture in developed countries.

Finally, although CSA reinforces the long-established focus on production growth, the evidence base to suggest that growth approaches are the solution to, and not just the cause of, the climate problem remains at best inconclusive.

ACKNOWLEDGEMENTS

Dominic Moran and Anita Wreford acknowledge financial support from the Scottish Government Rural and Environmental Science and Analytical Services division. Dominic Moran acknowledges funding from the UK Economic and Social Research Council ESRC Grant under grant number ES/N013255/1.

REFERENCES

Alcott, B. (2005), 'Jevons' paradox', *Ecological Economics*, **54** (1), 9–21.

Andersson, J.A. and S. D'Souza (2014), 'From adoption claims to understanding farmers and contexts: A literature review of conservation agriculture (CA) adoption among smallholder farmers in southern Africa', *Agriculture, Ecosystems and Environment*, **187**, 116–132.

Atela, J.O. (2012), *The Politics of Agricultural Carbon Finance: The Case of the Kenya Agricultural Carbon Project*, Steps Centre, http://steps-centre.org/wp-content/uploads/Carbon-Agriculture.pdf (accessed September 2013).

Branca, G., N. McCarthy, L. Lipper and M.C. Jolejole (2011), *Climate-Smart Agriculture: A Synthesis of Empirical Evidence of Food Security and Mitigation Benefits from Improved Cropland Management*, Mitigation of Agriculture in Climate Change Series 3, Rome: FAO.

Bromhead, M.-A., R. Sessa and S.J. Sadeghian (2013), 'Mainstreaming CSA into national policies and programmes', Module 13 in *Climate-Smart Agriculture Sourcebook*, Rome: FAO.

Civil Society (2015), 'Don't be fooled! Civil society says no to "climate-smart agriculture" and urges decision-makers to support agroecology', http://www.climatesmartagconcerns.info/cop21-statement.html (accessed October 2015).

Climate Focus (2011), 'Carbon market and climate finance for agriculture in developing countries', Prepared for Agriculture and Carbon Market Assessment, DFID Climate Change, Agriculture and Food Security Policy Research Programme, Department for International Development, Glasgow, UK.

Climate Summit (2014), *Agriculture: Global Alliance for Climate-Smart Agriculture Action Plan Provisional Copy*, New York: UN HQ, http://www.un.org/climatechange/summit/wp-content/uploads/sites/2/2014/09/AGRICULTURE-Action-Plan.pdf (accessed October 2014).

IPCC (2014), *Climate Change 2014: Synthesis Report: Contribution of Working Groups I, II and III to the Fifth Assessment Report of the Intergovernmental Panel on Climate Change*, ed. Core Writing Team, R.K. Pachauri and L.A. Meyer, Geneva: IPCC.

FAO (2012), *Towards the Future We Want: End Hunger and Make the Transition to Sustainable Agricultural and Food Systems*, Rome: FAO.

FAO (2013a), *Climate-Smart Agriculture Sourcebook*, Rome: FAO.

FAO (2013b), 'Farmers struggle to adopt climate-smart methods', News article, http://www.fao.org/news/story/en/item/181017/icode/ (accessed June 2014).

FAO (2014), *Mid-Term Evaluation of Climate Smart Agriculture: Capturing the Synergies between Mitigation, Adaptation and Food Security in Malawi, Vietnam and Zambia – GCP/INT/139/EC*, http://www.fao.org/fileadmin/user_upload/oed/docs/GCP%20INT139EC_2014_ER.pdf (accessed January 2015).

Gattinger, A., J. Jawtusch, A. Muller and P. Mäder (2011), *No Till Agriculture – A Climate-Smart Solution?*, Climate Change and Agriculture Report No. 2, Aachen, Germany: Bischöfliches Hilfswerk Misereor.

Giller, K.E., E. Witter, M. Corbeels and P. Tittonell (2009), 'Conservation agriculture and smallholder farming in Africa: The heretics' view', *Field Crops Research*, **114**, 23–34.

Godfray, H.C.J., J.R. Beddington, I.R. Crute, L. Haddad, D. Lawrence, J.F. Muir, J. Pretty, S. Robinson et al. (2010), 'Food security: The challenge of feeding 9 billion people', *Science*, **327**, 812–818.

GSI (2010), 'Measuring irrigation subsidies: Policy recommendations from a Spanish case study', Policy brief, Global Studies Initiative, International Institute for Sustainable Development, Geneva, www.globalsubsidies.org.

Hilger, T., A. Keil, M. Lippe, M. Panomtaranichagul, C. Saint-Macary, M. Zeller, W. Pansak, T.V. Dinh et al. (2013), 'Soil conservation on sloping land: Technical options and adoption constraints', in H.L. Fröhlich, P. Schreinemachers, K. Stahr and G. Clemens (eds), *Sustainable Land Use and Rural Development in Southeast Asia: Innovations and Policies for Mountainous Areas*, Berlin: Springer, pp. 229–279.

Ito, M., T. Matsumoto and M. Quinones (2007), 'Conservation tillage practice in sub-Saharan Africa: The experience of Sasakawa Global 2000', *Crop Protection*, **26**, 417–423.

Kaczan, D., A. Arslan and L. Lipper (2013), *Climate-Smart Agriculture? A Review of Current Practice of Agroforestry and Conservation Agriculture in Malawi and Zambia*, ESA Working Paper No. 13-07, Rome: Agricultural Development Economics Division, Food and Agriculture Organization of the United Nations.

Lang, C. (2011), 'Bad for the climate, risky for farmers, good for the World Bank: Andrew Steer's proposal for "success in Durban"', *REDD-Monitor*, http://www.redd-monitor.org/2011/11/22/bad-for-the-climate-risky-for-farmers-good-for-the-world-bank-andrew-steers-proposal-for-success-in-durban/ (accessed September 2013).
Lee, H.-C., B.A. McCarl, U.A. Schneider and C.-C. Chen (2006), 'Leakage and comparative advantage implications of agricultural participation in greenhouse gas emission mitigation', *Mitigation and Adaptation Strategies for Global Change*, **12** (4), 471–494.
Lilliston, B. (2015), 'The clever ambiguity of climate smart agriculture', Institute for Agriculture and Trade Policy blog post, http://www.iatp.org/blog/201512/the-clever-ambiguity-of-climate-smart-agriculture.
Lipper, L., W. Mann, A. Meybeck and S. Sessa (2010), *Climate-Smart Agriculture: Policies, Practices and Financing for Food Security, Adaptation and Mitigation*, Rome: FAO.
Mbow, C., M. Van Noordwijk, E. Luedeling, H. Neufeldt, P.A. Minang and G. Kowero (2014), 'Agroforestry solutions to address food security and climate change challenges in Africa', *Current Opinion in Environmental Sustainability*, **6**, 61–67.
McCarthy, N. and J. Brubaker (2014), *Climate-Smart Agriculture and Resource Tenure in Sub-Saharan Africa: A Conceptual Framework*, Rome: FAO.
Meybeck, A. and V. Gitz (2012), *Greening the Economy with Climate-Smart Agriculture*, Prepared as technical input for the Hanoi Conference on Agriculture, Food Security and Climate Change, 3–7 September, Rome: FAO.
Moran, D., M. MacLeod, E. Wall, V. Eory, A. McVittie, A. Barnes, R. Rees, C.F.E. Topp et al. (2011), 'Marginal abatement cost curves for UK agricultural greenhouse gas emissions', *Journal of Agricultural Economics*, **62** (1), 93–118.
Nelson, G.M., W. Rosegrant, J. Koo, R. Robertson, T. Sulser, T. Zhu, C. Ringler, S. Msangi et al. (2009), *Climate Change Impact on Agriculture and Costs of Adaptation*, IFPRI, http://www.ifpri.org/sites/default/files/publications/pr21.pdf.
New Zealand Agricultural Greenhouse Gas Research Centre (2014), 'Reducing New Zealand's agricultural greenhouse gas emissions: How we are getting there', www.nzagrc.org.nz/fact-sheets.htm.
Oppenheimer, M., M. Campos, R. Warren, J. Birkmann, G. Luber, B. O'Neill and K. Takahashi (2014), 'Emergent risks and key vulnerabilities', in C.B. Field, V.R. Barros, D.J. Dokken, K.J. Mach, M.D. Mastrandrea, T.E. Bilir, M. Chatterjee, K.L. Ebi et al. (eds), *Climate Change 2014: Impacts, Adaptation, and Vulnerability, Part A: Global and Sectoral Aspects: Contribution of Working Group II to the Fifth Assessment Report of the Intergovernmental Panel on Climate Change*, Cambridge, UK: Cambridge University Press, pp. 1039–1099.
Osorio, C.G., D.W. Abriningrum, E.B. Armas and M. Firdaus (2011), *Who Is Benefiting from Fertiliser Subsidies in Indonesia?*, World Bank Policy Research Working Paper 5758, Jakarta: Poverty Reduction and Economic Management Unit, World Bank, East Asia and Pacific Region.
Pervin, M., S. Sultana, A. Phirum, I.F. Camara, V.M. Nzau, V. Phonnasane, P. Khounsy, N. Kaur et al. (2013), *A Framework for Mainstreaming Climate Resilience into Development Planning*, IIED Working Paper, http://pubs.iied.org/10050IIED (accessed June 2014).
Pye-Smith, C. (2011), *Farming's Climate-Smart Future: Placing Agriculture at*

the Heart of Climate Change Policy, Wageningen, the Netherlands: CTA and Copenhagen: CGIAR Research Program on Climate Change, Agriculture and Food Security (CCAFS).
Richards, M., T. Sapkota, C. Stirling, C. Thierfelder, N. Verhulst, T. Friedrich and J. Kienzle (2014), *Conservation Agriculture: Implementation Guidance for Policymakers and Investors*, Climate-Smart Agriculture Practice Brief, Rome: FAO, http://www.fao.org/3/a-i4066e.pdf (accessed February 2015).
Shames, S., E. Wollenberg, L.E. Buck, P. Kristjanson, M. Masiga and B. Biryahaho (2012), *Institutional Innovations in African Smallholder Carbon Projects*, CCAFS Report No. 8, Copenhagen: CGIAR Research Program on Climate Change, Agriculture and Food Security (CCAFS), www.ccafs.cgiar.org (accessed August 2013).
Sharma, S. and S. Suppan (2011), *Elusive Promises of the Kenya Agricultural Carbon Project*, Minneapolis, MN: Institute for Agriculture and Trade Policy (IATP).
Steenwerth, K.L., A.K. Hodson, A.J. Bloom, M.R. Carter, A. Cattaneo, C.J. Chartres, J.L. Hatfield, K. Henry et al. (2014), 'Climate-smart agriculture global research agenda: Scientific basis for action', *Agriculture and Food Security*, **3** (11), http://www.agricultureandfoodsecurity.com/content/3/1/11 (accessed January 2015).
Thierfelder, C., S. Cheesman and L. Rusinamhodzi (2012), 'A comparative analysis of conservation agriculture systems: Benefits and challenges of rotations and intercropping in Zimbabwe', *Field Crops Research*, **137**, 237–250.
Wang, W., F. Koslowski, D.R. Nayak, P. Smith, E. Saetnan, X. Ju, L. Guo, G. Han et al. (2014), 'Greenhouse gas mitigation in Chinese agriculture: Distinguishing technical and economic potentials', *Global Environmental Change*, **26** (May), 53–62.
Westermann, O., P. Thornton and W. Förch (2015), *Reaching More Farmers – Innovative Approaches to Scaling Up Climate Smart Agriculture*, CCAFS Working Paper No. 135, Copenhagen: CGIAR Research Program on Climate Change, Agriculture and Food Security (CCAFS).
World Bank (2012–2014), *Kenya Agricultural Carbon Project*, http://www.worldbank.org/projects/P107798/kenya-agricultural-carbon-project?lang=en (accessed February 2015).

8. Global warming and changes in marine ecosystems: economic consequences and adjustment issues

Clem Tisdell

INTRODUCTION

Marine areas supply humanity with many valuable ecosystem services. Costanza et al. (2014) estimated the value of ecosystem services provided by marine areas to be US$49.7 trillion in 2011 and point out that this is a much lower aggregate value than in 1997. Their conclusion seems to be that this reduced value is due to deterioration in marine ecosystems. Marine services consist of marketed, partially marketed and unmarketed ones. For example, for the most part, the provision of marine fish is marketed, marine tourism and recreational services are partly marketed, and the value of marine areas in moderating rising atmospheric temperatures, caused by elevated levels of greenhouse gases (GHG), is unmarketed. However, rising atmospheric levels of GHG, particularly CO_2, due to well-known anthropogenic factors, are throwing the supply of marine ecoservices into disequilibrium. As a result, the economic value and supply of these services are expected to alter drastically and spatially in the coming decades. Most communities which depend on marine resources for their livelihood will need to adjust to these changing conditions. Governments will need to consider whether and when they should intervene to enable affected communities to adapt to altered marine environments or, if possible, to select policies to retard the pace of anticipated adverse changes in these environments.

This chapter unfolds as follows. First, it outlines major changes in marine environments which natural scientists expect to occur as a result of elevated levels of GHG in the atmosphere and considers their general implications for the presence of living organisms and for the supply of marine ecoservices. Second, attention is given to insights obtained from a Norwegian investigation of the economic impacts of GHG-induced changes in marine ecosystems. Third, GHG-induced losses of coral reefs and their potential economic impacts are discussed. Proposed policies for responding to predicted changes in marine ecosystems are then examined, and this is followed by a discussion of resilience.

Why is this topic important? Globally marine ecosystems have been estimated to provide a very large percentage of the flow of the economic value of ecosystem services. Based on the estimates of Costanza et al. (2014), it can be deduced that marine ecosystems in 2011 accounted for 39% (US$49.7 trillion) of the aggregate economic value of the flow of ecosystem services. However, if the estimated flow of the value of ecosystem services from tidal marshes and mangroves is added to these figures, the total value of the flow of marine ecosystems in 2011 increases to US$74.5 trillion, 59.7% of the global flow. Aggregate human well-being could be severely reduced as a result of losses in these services because of climate change. Furthermore, the potential distributional consequences of changes in marine ecosystems as a result of climate change are important because many communities depend (particularly in developing countries) heavily on marine resources for their welfare. The topic is also of significance because the discussion of it illustrates the challenges we face in developing adequate economic and ecological models to predict changes in the provision of marine ecosystem services and in their economic values as a result of climate changes. Moreover, this chapter highlights the inadequacies (from both an economics and an ecological point of view) of assessing changes in the provision of ecosystem services and in the economic value of examining changes in ecosystems in isolation from one another.

ABIOTIC CHANGES IN MARINE ECOSYSTEMS DUE TO ELEVATED GHG LEVELS

Elevated levels of GHG are causing global warming of the atmosphere (mainly as a result of increased levels of CO_2) and associated physical and chemical changes in marine environments. In turn, this is altering marine ecosystems and the populations of living organisms they support. Consequently, the supply of marine ecosystem services is changing and is expected to continue to do so as atmospheric GHG levels rise. Many human communities will be worse off as a result even though some may gain economically. Predictions about the economic and social impacts of such changes are vitally dependent on the prognosis of natural scientists about the abiotic and biotic changes to be expected as a result of climate change. In order to assess their economic impacts and to identify rational policy responses, the nature, geographical location and timing of changes in marine ecosystems attributable to GHG emissions (and other causes)

Table 8.1 An indication of the heterogeneity of marine ecosystems

Ecosystem identifiers	Sets of ecosystems
In coastal locations	Salt marshes, mangroves, estuaries.
On rocky substrates	Rocky intertidal, kelp forests, coral reefs.
On soft substrates	Sandy shores, seagrass meadows, shelf sea benthos.
Vast oceanic ecosystems	Pelagic, polar and ice-edge, deep sea.

Source: Based on Brierley and Kingsford (2009: R606, table 2).

need to be identified. This is a complex and formidable task. Therefore, it is not surprising that most scientific opinions about the dynamics and spatial aspects of changes in marine ecosystem change are subject to significant degrees of uncertainty. This results inevitably in economic assessments reliant on those informed opinions being subject to uncertainty: uncertainty which is compounded (as a rule) by shortcomings in economic data and analysis. While this does not mean that it is impossible to make useful economic assessments of the cost–benefit and economic impact issues raised by GHG-induced changes in marine ecosystems, it is important to keep in mind their likely limitations.

The first challenging aspect which might be noted is the considerable variety of marine ecosystems. Table 8.1 provides an initial indication of that variety. However, in practice, types of ecosystems are much more heterogeneous than indicated. For example, the type of environmental conditions experienced by coral reefs can be expected to vary geographically. This results in differences in the community of living organisms which they support. Furthermore, the functioning of many of these ecosystems is interdependent, and therefore they are not independent entities.

A variety of changes in marine environments are already occurring or are confidently expected to occur as a result of elevated GHG levels in the atmosphere. These include:

- Increasing *acidification* of marine waters induced by higher levels of atmospheric CO_2. This is expected to have adverse consequences for the survival and growth of calciferous organisms, particularly molluscs and hard corals and, to a lesser extent, crustaceans (Doney et al., 2009; Kroeker et al., 2010).
- The area of the ocean suffering serious *oxygen depletion* is predicted to increase (Stramma et al., 2012). Oxygen deficient areas become virtual dead zones and support few, if any, organisms of value to human beings.

- *Sea level rises* are expected as a result of global warming. This is because this warming expands the volume of water in the oceans and also melts global ice cover, thereby adding to this volume. The effects on marine ecosystems depend on the speed of sea level rise; the faster and more pronounced the rise, the more difficult it is for existing ecosystems to adjust (Solomon et al., 2007: chap. 5). Some coral reefs are likely to be drowned, and dislocation can be expected in mangroves, coastal wetlands and estuaries.
- *Changes* seem to be likely in ocean circulation (currents), upwelling (vertical movements of sea water) and the vertical (heat) stratification of oceans (Sumaila et al., 2011). Among other things, this is likely to affect the local availability of fish (see, for example, Garza-Gil et al., 2011).
- *Rising ocean temperatures* affect the distribution of marine species (see, for example, Dulvy et al., 2008). These rising temperatures can result in coral bleaching, the overgrowth of corals by algae and a likely reduction in their areal range. The abundance of fish and the presence of different species of fish spatially will also alter significantly (Cheung et al., 2010).
- *Increasing storm frequency and severity* in some regions (Solomon et al., 2007: chap. 5) are expected to have adverse economic consequences. For example, mariculture is likely to be adversely impacted by increased damage to fish pens and marine-based installations, and seaweed beds are more prone to destruction. Furthermore, erosion of coral reefs is likely to accelerate, thereby reducing the supply of this valuable natural resource.

All of the above-mentioned abiotic changes to existing marine ecosystems will alter their biotic components. As a result, many existing marine ecosystems (possibly most) will be replaced by new ecosystems. There will be a large-scale geographical redistribution of the maximum potential fisheries catch (Cheung et al., 2010) and alterations in the geographical distribution of marine species. However, the changes will not be instantaneous, and the dynamic pathways which existing ecosystems will follow in their transition are likely to vary, with diverse economic consequences. Predicting their economic consequences is difficult because ecologists are uncertain about the dynamics of the adjustment of marine ecosystems in response to climate change (Miller et al., 2010). However, present indications are that some countries will benefit economically from such changes, for instance Norway. Others are likely to be disadvantaged (Allison et al., 2009). Modelling by Armstrong et al. (2012) suggests that Norway will make economic gains as a result of an increase in available fish stocks. On the other hand, loss of coral reefs may disadvantage Australia and many lower income countries economically. Let us consider the economic modelling of Armstrong et al. (2012) for Norway and then discuss the possible economic consequences of loss of coral reefs due to climate change, ocean acidification and other factors.

THE ECONOMIC IMPACT OF CLIMATE CHANGE AND OCEAN ACIDIFICATION ON NORWEGIAN FISHERIES

Modelling by Armstrong et al.

Armstrong et al. (2012) undertook the heroic task of estimating the economic impact of ocean acidification and climate change on the aggregate gross revenues generated by the Norwegian fisheries and mariculture. They estimate this for the 100-year period 2010–2110. Because they assume that the real prices which prevailed for Norwegian fish in the period 2001–2010 will continue to prevail in the period 2010–2110, their results principally depend on anticipated changes in the volume of fishing output during the focal 100-year period. Their findings rely primarily on the results of the meta-analysis conducted by Kroeker et al. (2010) of the effects of ocean acidification on marine organisms and to a lesser extent on the estimates of Hendriks et al. (2010). A problem of depending on such meta-estimates is that these estimates do not relate to particular regions and are aggregative in nature. They are, for example, not specific to the Norwegian exclusive economic zone (EEZ). Moreover, the estimates of changes in survival, growth and calcification of species relate to broad groupings of species, for example 'all' fish, molluscs and crustaceans are considered as separate groups (see Kroeker et al., 2010: 1426). This ignores any variability within categories.

Armstrong et al. (2012) assume that the composition of fish species in the Norwegian EEZ does not alter over their 100-year period. However, this is unlikely to be so. Fish species appear, for example, to be altering their geographic location as a result of global warming and other effects of climate change (Cheung et al., 2010). Furthermore, the dynamics of changes in fish abundance for a 100-year period are far from certain: it may not accord with the linear relationship assumed by Armstrong et al. in their analysis.

In their economic analysis, Armstrong et al. limit their attention to changes in the supply of provisioning services (food) expected to occur as a result of the effects of ocean acidification on Norwegian marine fish and aquaculture. They estimate the changes in the gross revenue which will be obtained when the revenue streams alter for fish, molluscs and crustaceans separately and for these combined. These estimates can be regarded (in terms of economic impact analysis) as variations in the level of first-round monetary injections into the Norwegian economy. Hence, they represent a form of economic activity analysis rather than cost–benefit analysis. Estimates for best and worst case scenarios are provided for fish, molluscs and crustaceans as well as for the aggregates of their changed revenue streams. Armstrong et al. (2012) estimate that the combined total change in revenue for the period 2010–2110 in the best scenario case is an increase of 105 059 million 2010-NOK in revenue and in the worst case, a decrease in revenue of 3926 million 2010-NOK. (At the end of June 2010, a Norwegian krone equalled US$0.152262[1].)

Allowing for Uncertainty and the Discounting of Economic Values

It could be argued on the basis of Laplace's principle of insufficient reason that these end-values and all those values in between are equally probable. In this case, the probability distribution of changes in total revenue is rectangular, and the expected change in the gross revenue obtained in Norwegian fisheries and aquaculture is an increase of 50 566.6 million 2010-NOK for the period 2010–2110. Furthermore, given this probability distribution, the likelihood (0.06) of a negative effect is very low. Therefore, it seems highly likely (probability 0.94) that, in aggregate, Norwegian fisheries and aquaculture will generate increased revenue as a result of ocean acidification.

Armstrong et al. (2012) also provide discounted values for these changed revenue streams, but they do not explain their rationale for doing this. It is unclear in this context whether the discounting is assumed to indicate time preference in relation to the current generations or the coefficient of concern of current generations for future generations.

The Norwegian results are heavily influenced by the fact that the bulk of Norway's marine food production and revenue is obtained from fish, with only a small proportion being obtained from molluscs and crustaceans. Moreover, if Laplace's principle is adopted, their estimates reveal that the expected changes in revenue from mollusc and crustacean production are both negative, but small in relation to the expected increase in revenue from fish production. Nevertheless, given Laplace's principle of insufficient reason, it is highly probable (probability 0.80) that revenue generated by mollusc production will fall. The likelihood of a negative change in revenue from crustacean production is less: 0.55.

Note that, even if the results of Armstrong et al. are reliable for Norway, it cannot be concluded that all countries will obtain increased revenue from marine food production as a result of ocean acidification. For example, in Australia's case, crustacean and mollusc production accounts for a high proportion of the revenue generated by its marine food production. Furthermore, it is highly likely that Australia's coral reef-based fisheries will be adversely affected. Moreover, coral reef deterioration as a result of global climate change may have an adverse impact on Australia's revenues generated by tourism.

The Validity of Ecological Assumptions

A problem in undertaking economic analysis of this type is that ecologists are uncertain about the impacts of climate change on the biological composition of ecosystems and the dynamics of changes in these. For example, according to the modelling of Cooley et al. (2012), the lead times before mollusc productivity is adversely affected by climate change could be very long and are predicted to vary considerably between countries. They predict that virtually no reduction in mollusc production will occur until a 'transition decade' is reached in which aragonite saturation of sea water begins to fall below a critical level. Aragonite is a form of calcium carbonate and is used by molluscs to produce their shells. Taking 2010 as the base year, their lead times vary from a minimum of 14 years for Ecuador to a maximum of 45 years in North Korea. This implies that Ecuador's transition decade would begin in 2024, whereas that of North Korea would not start until 2055. The lead time is the length of time between 2010 and the year in which aragonite saturation is predicted to fall below a critical level. The lead times predicted for Australia and the USA are respectively 17 and 18 years and imply that their transition decade would start in 2027 and 2028 respectively.

How accurate these predictions are is difficult to tell. They indicate, however, that some countries would have to prepare for a reduction in their mollusc production at an earlier stage than others. Furthermore, the nature of the impact on the productivity of molluscs predicted by Cooley et al. (2012) differs from that assumed in the analysis of Armstrong et al. (2012) because the analysis of Armstrong et al. is based on the prediction that mollusc productivity declines at a constant rate from 2010 onwards.

An additional limitation of current ecological knowledge about the consequences for marine ecosystems of GHG emissions is that it is based on partial analysis. For example, the consequences of GHG emissions for marine organisms, including their location, depend on multiple effects (listed in the previous section) (see, for example, Ainsworth et al., 2011). Ocean acidification is just one of these effects. For the purposes of economic evaluation, the combined multiple effects of GHG emissions need to be taken into account, not just one. For example, in the Norwegian case, what is the likelihood that fish species in its EEZ will alter, and what is the probability that the primary production of available food for fish (see Ainsworth et al., 2011) will increase or decrease as a result of changing ocean currents or the increased rate at which ice sheets in the Arctic are melting? None of these factors are allowed for in the analysis by Armstrong et al. (2012).

The discussion of the Norwegian case study of the consequences of GHG emissions for marine productivity should alert us to the following: some regions and marine-based industries are likely to make economic gains, whereas others will experience economic losses as a result of global environmental changes induced by rising levels of GHG. Furthermore, the state of ecological modelling makes it very difficult to predict confidently the timing and spatial patterns of change in marine productivity as a result of global warming. This compounds the difficulty of obtaining reliable estimates of the economic impact of global climate change. In addition, different economic and ecological modelling by Eide (2007) and by Lam et al. (2014) indicate that the economic benefits from changes in marine fisheries in the Arctic are likely to be lower than the study by Armstrong et al. (2012) suggests.

CORAL REEFS AND CLIMATE CHANGE

Costanza et al.'s View of the Changing Economic Value of Coral Reefs Due to Climate Change

Costanza et al. (2014: table 3) argue that the value of ecosystem services per hectare provided in 2011 by coral reefs was by far the highest for any biome. They also supply data to support their view that this biome recorded the greatest reduction in the annual economic global value of ecosystem services provided by any biome comparing this supply in 2011 to that in 1997. This is so if their hypotheses are accepted that: (1) the global area of coral reefs more than halved in this period (and that their vacated areas were replaced by seagrass and algal beds); and (2) the estimated real price per hectare of coral reefs in 2011 provides a more suitable basis for estimating the change in the total global value of this biome than does the use of 1997 real unit values. Their method involves valuing ecosystem services in both 1997 and 2011 by either the real unit values prevailing in 1997 or those prevailing in 2011. Dollar values for 1997 and 2011 are adjusted to the value of the US dollar in 2007. Note that they attribute the reduction in the global economic value of coral reefs mainly to a reduction in their area as a result of coral bleaching caused by global warming.

They find on the above-mentioned basis that there was a reduction in the total value of coral reef ecosystem services provided in 2011 compared to 1997 of US$10.9 trillion. This was only offset to a relatively small extent by the estimated increase in the total value of seagrass and algal biomes. Although the real value per hectare of ecosystem services provided by seagrass and algal biomes was lower than for coral areas, their total real economic value rose by US$1.0 trillion per year. This was because of their increased areal extent based on the assumption that they replaced depleted coral reef areas. Consequently, the decline in the net amount of the aggregate annual economic value of ecosystem services attributed to reduced coral cover is US$9.9 trillion, given the estimates of Costanza et al. (2014).

A Different Estimate to That of Costanza et al. of the Changing Economic Value of Coral Reefs Due to Climate Change

The way in which Costanza et al. (2014) estimate changes in the aggregate value of ecosystem services provided by biomes is heavily influenced by their choice of whether to determine those based on 1997 real values per hectare or to do so based on 2011 real values per hectare. They consider whether they should apply 1997 real values to both of their focal years or 2011 values in order to estimate these changes. They favour the latter. They state: 'Given the more comprehensive unit values employed in the 2011 estimates, this is [these give] a better approximation than using the 1997 unit values, but certainly still are conservative estimates' (Costanza et al., 2014: 156). However, they do not address the question of whether it would be even more appropriate to use 1997 real unit values for valuing the ecosystem services provided in 1997 and the real unit values prevailing in 2011 for valuing those services supplied in 2011. If this procedure is used, different conclusions follow from those of Costanza et al. Given this alternative approach, the aggregate economic value of the ecosystems services generated by coral reefs in 2011 is much higher than in 1997, when these values are expressed in terms of 2007 US dollars. This is so despite the area of coral reefs being more than halved. This is because the increased value of coral areas' ecosystem services per unit of their remaining area more than compensates for the reduction in these areas, given the data used by Costanza et al. This result is plausible on economic grounds, but depends on how economic value is measured.

Costanza et al. estimated that, based on 2007 real values, the value of ecosystem services provided per hectare by coral reefs was $8364 in 1997, but in 2011 this rose to $352 249. The 2011 figure is 42 times that for 1997.

On the other hand, it was estimated that the global area covered by coral reefs in 2011 was 0.45 of that in 1997. The relative decrease in this area is far outweighed by the comparative increase in the estimated value of coral reefs per hectare. Consequently, according to these estimates, coral reefs yielded annual ecosystem services of much greater aggregated global value in 2011 than in 1997. How could this be so? One possibility is that the data are faulty, but there are also other possible economic reasons for this result. Consider some of these.

Explaining the Increased Economic Value of Coral Reefs Despite Their Reduced Area Due to Global Warming

Two different and often conflicting approaches to economic valuations can be found in the literature. One is based on economic activity or impact analysis, and the other relies on traditional economic welfare analysis. Although legitimate doubts exist about whether economic activity analysis should be regarded as a valuation technique, it is often applied as if it is, and of course it does have economic implications. While some environmental losses can increase aggregate expenditure obtained by industries reliant on affected environments, at the same time economic welfare (based on traditional economic welfare analysis) can fall (Tisdell, 2012). Take the loss of coral reefs as an example.

The value per hectare both of marketed commodities and of non-marketed ones associated with coral reefs probably increased between 1997 and 2011. In the case of marketed commodities two factors may have been at work. The demand curve for important commodities associated with coral reefs most likely shifted upwards, for example particularly the demand for tourism services. Second, given normal demand curves, the reduced global area of reefs would have increased their scarcity value and forced up the prices of marketed commodities which are complements of reef availability. If the demand schedules for these commodities are mainly in their inelastic range, total revenue associated with the dwindling global area coral reefs will increase, as illustrated in Tisdell (2015a).

Furthermore, estimates of the non-marketed economic values per hectare of remaining reefs may rise as the total global area of coral reefs declines. Again, two factors can contribute to this: the sum of marginal valuations of the supply of non-marketed goods associated with the increased area of coral reefs can be expected to decline with an increase in their global area; and the collective marginal evaluation of this supply may shift upwards, as explained in Tisdell (2015a). Particularly in developed countries, the non-marketed economic values of coral reefs are likely to exceed their marketed values (Stoeckl et al., 2014).

Costanza et al. (2014) claim that it is desirable (in order to underline the value of natural ecosystems) to estimate the expenditure which would be generated by non-marketed goods if they could be marketed, and include this in estimates of their economic value. This is a type of 'proxy' expenditure approach, as explained in Tisdell (2015b: chap. 16).

In all the above circumstances, the expenditure approach (whether based on only actual expenditures or on those plus 'proxy' expenditures) to valuing the availability of coral reefs can result in their total estimated economic value rising as their global availability declines. Consequently, those communities fortunate enough to have some areas of surviving coral reefs can have increased incomes from their available reefs. Nevertheless, using standard economic welfare analysis, global economic welfare is likely to be reduced.

Aspects of this global impoverishment are illustrated in Tisdell (2015a). However, in undertaking this evaluation, it cannot be assumed that areas which lose their coral reef cover will become completely devoid of economic value. In Costanza et al.'s modelling, it is assumed that areas which become devoid of coral reef cover are replaced by seagrass and that this has a much lower economic value per hectare than coral reef cover.

It is also worth noting that per-hectare economic values of ecosystems estimated by Costanza et al. (2014) and by de Groot et al. (2012) are the averages of marginal values. They are based on meta-values calculated for selected (limited) reef areas. This raises two potential problems. Possibly on the whole, these economic values have only been estimated for the most valuable reefs. If so, the estimates will overstate values per hectare for coral reefs in aggregate. This is a potential benefit transfer problem (Rolfe and Windle, 2012). A second problem arises given that the values used in these studies are marginal values, because mean marginal values are not mean average aggregate values. Multiplying marginal values per hectare of an ecosystem by its total area will only provide a reliable guide to its aggregate economic value if marginal and average values of the ecosystem are a constant function of its area and equal. This is effectively assumed in the above-mentioned studies. However, the above economic analysis implies that this assumption is unlikely to be satisfied. Furthermore, Costanza et al. (2014) suppose that coral reef areas (and other ecosystem areas) which were lost in the period between 1997 and 2011 transited entirely from one type of ecosystem to another. However, this assumption is ecologically suspect.

Biological Resilience of Coral Reefs

Scientific evidence that some coral reefs can show considerable ability to recover from coral bleaching events should not be ignored, even though the diversity of their coral assemblages tends to be reduced following such events (Adjeroud et al., 2009; Mumby et al., 2011; Roff et al., 2014). We cannot rely on comparative static analysis to assess satisfactorily the economic consequences of changes in marine ecosystems. The trajectories (dynamic paths) of ecosystem change must be accounted for, taking into account, for example, the type of issues raised by Hoegh-Guldberg et al. (2007). Several of these issues are discussed in Tisdell (2015a). The analysis of Costanza et al. (2014) does not pay enough attention to the dynamics of ecosystem change, because it only makes comparisons between two years. Furthermore, they suppose (at odds with the ecological findings mentioned above) that coral reef areas damaged by global warming transited entirely to seagrass ecosystems. Consequently, economic resilience issues are not considered.

ECONOMIC POLICIES FOR RESPONDING TO CLIMATE-INDUCED CHANGES IN MARINE ECOSYSTEMS

Now consider economic policies which may be adopted for responding to climate change. Two types of policies can be considered:

- policies to retard ecosystem change (mitigation strategies);
- policies which are based on the inevitability of ecosystem change and which are designed to cope with it (adjustment strategies).

In some cases, these policies may be pursued in conjunction. For example, ecosystem change may be retarded, but it may still be necessary to adopt policies for economic and social adjustment to the changing ecosystem. Let us consider economic aspects of the first-mentioned type of policy response initially and then pay attention to the second type of policy response.

Economic Aspects of Policies to Slow Changes in Marine Ecosystems

It is generally accepted that, even if there are stresses on marine ecosystems caused by global climate changes, in some cases it is possible to retard changes in these systems by reducing local anthropogenic stresses. The possibility of doing this depends on locational factors and the type of marine ecosystem being placed under stress. The deterioration of coral reefs in some locations may, for example, be retarded by reducing the runoff of nutrients and sediment from nearby land areas where runoff is a function of anthropogenic activity. This is a policy option being pursued for helping to conserve parts of the Great Barrier Reef. If, for example, fertilizer use in agriculture is identified as an important contributor (via nutrient runoff) to coral reef stress, then from an economics point of view its use should be reduced until the marginal economic benefit of improved reef quality equals the marginal loss in economic benefit from agriculture.[2]

In countries such as Australia, where coral reefs are spread geographically over very large areas, it is unlikely to be economic to adopt spatially uniform measures to conserve reef areas affected by human activity. Appropriate measures will depend on relative economic values. For example, if in the most northerly region of the Great Barrier Reef the economic value of intensified agriculture is low relative to the loss in economic value it causes to coral reefs, this weakens the case for encouraging the intensification of agriculture in this northern area.

The above analysis may also be applied to changes in other marine ecosystems. For example, increases in nutrient loads and sediments in the China Sea contribute to the increased frequency and the extent of red tides in this area, and these have adverse impacts on the fisheries located there. Therefore, China needs to consider in what areas and to what extent priority should be given to reducing nutrient and sediment discharges into the China Sea.

Spatial issues of the above kind have received little attention from economists. However, given that it is unlikely to be economic to conserve existing marine ecosystems everywhere, they are important. There is no escaping the fact that it is desirable to consider economic trade-offs in making decisions about the conservation of ecosystems.

Adjustment Policies Based on the Assumption that Changes in Marine Ecosystems Are Inevitable

Despite attempts to counter changes in marine ecosystems attributable to climate change, some changes seem to be inevitable. That raises the question of what policies, if any, should be adopted by governments to respond to the elements of climate change which they do not individually control. This is a very complex matter given that the timing of ecosystem changes due to climate change and their nature are still subject to considerable uncertainty.

Views about the extent to which governments should intervene in adjustments to climate change can vary substantially. On the one hand, the view may be adopted that, at the very most, only limited government intervention is called for. Particularly if environmental change is slow, communities may adjust adequately to it of their own accord. At the other end of the spectrum is the belief that governments should play a leading role in preparing individuals and communities for such changes and mandate that particular precautionary actions be taken.

The timing of adjustment to climate change can be problematic. In terms of timing, some adjustments to climate change may be premature and in other cases not made early enough. For example, there may be projects for which the length of their economic life is such that, if they are proceeded with now, they will give a sufficient return to make them worthwhile before the full impact of climate change adversely affects them. However, at a later time, these projects may no longer be economic given the increased immediacy of significant adverse environmental impacts from climate change. For instance, some types of mariculture may no longer be economic once impacts from climate change become significant. Prior to this they may still be economic even if some impacts from climate change are apparent.

The question of the extent to which governments should prepare communities for climate change and direct their responses is complex. Those who favour minimal government intervention in the economy might argue that the government ought to limit itself to providing public goods to facilitate adjustment to climate change and other commodities that cannot be adequately supplied by markets. The supply of public goods could include the supply of information about the expected nature and timing of climate change. However, there is the problem that information provided by governments is not always reliable. Furthermore, there may be a case for governments undertaking works to provide local public goods which delay some consequences of climate change, for example in some cases building or raising dykes to respond to sea level rises.

To some extent, insurance markets can be expected to respond to changes in the risks associated with climate change. For example, insurance premiums for mariculture operations facing increasing risks as a result of climate change can be expected to rise. This should send a market signal to mariculturalists that adjustments are called for.

RESILIENCE AND ALTERATIONS TO MARINE ECOSYSTEMS CAUSED BY GLOBAL CLIMATE CHANGE

Economic resilience in response to climate change has become a subject of increasing interest. However, the term 'resilience' is frequently used without paying much attention to its meaning. It usually refers to the ability of an entity to recover or bounce back to a previous state (or approach it) following dislocation by an adverse event. Basically, it is a dynamic concept.

Two aspects of resilience are relevant to the present discussion:

- the ability of marine ecosystems to recover from the adverse impacts of climate change; and
- the ability of socio-economic systems to do likewise.

A limitation of the concept in this context is that it tends to indicate implicitly that the biological and economic consequences of climate change are always adverse. However, as the cases considered in this chapter demonstrate, this is unlikely to be so in all localities.

The concept of resilience has achieved prominence because it has come to be regarded as the most important feature of the sustainability of ecological and economic systems. In reality, however, it is only one attribute of such systems which needs to be considered when assessing their performance if they are subjected to environmental changes, such as climate change (Tisdell, 2015a). A few relevant aspects of resilience can be mentioned here.

First, economic resilience can be increased by being prepared by taking precautionary and anticipatory actions. However, this can be done too early from an economics point of view. For example, although ocean acidification seems likely to eventually have adverse consequences on the cultivation of molluscs (such as oysters), a significant amount of time may expire before this effect occurs or becomes an economic concern (Cooley et al., 2012). In many cases, no changes in current practices will be economic for some time to come.

Second, the need for economic resilience in relation to changes in marine ecosystems depends to a large extent on the resilience of these systems or, more generally, the dynamics of their responses to climate change. Some marine ecosystems in some situations may exhibit a limited amount of resilience. There is, for example, evidence of partial recovery of coral reefs from heat stress (Roff et al., 2014). However, the effect of stresses on marine ecosystems is complex. It is likely to depend on several dynamic features, for instance whether the stresses are sudden, gradual, intermittent, and rising in magnitude. Also, the combination of stressors is important.

Despite all these difficulties of scientific prediction, current evidence indicates that there is a high probability that countries in low latitudes (nearest the equator) will suffer the most adverse economic consequences from changes in marine ecosystems due to climate change, whereas those in higher latitudes are likely to gain (Allison et al., 2009; Cheung et al., 2010; Sumaila et al., 2011). Consequently, tropical less developed countries (which depend heavily on living marine resources) are likely to be the most adversely affected economically by these changes. This is especially so for most island countries, and the capacity of these countries to cope with these changes is extremely limited (Tisdell, 2008). In fact, compared to most higher income countries, most lower income countries have limited scope and capacity to adjust to major threats to the livelihoods of their citizens which are expected as a result of anticipated alterations to their marine environments as a result of climate change. In other words, their scope and capacity for economic resilience in these circumstances are severely constrained. In many cases, substantial environmental deterioration is likely to trigger increased migration from less developed countries, and many island countries may no longer be inhabitable (Tisdell, 2008).

CONCLUDING COMMENT

The discussion in this chapter supports the view of Stoeckl et al. (2011: 126) that, in order to devise appropriate economic policies for the management of ecosystems, 'we do not just need information about total and marginal values [of the services they provide] but we also need information on the social, temporal and spatial distribution of these values'. This is especially important in the case of ecosystems which change their nature in response to alterations in environmental conditions, such as those associated with elevated levels of atmospheric GHG. It is also desirable that these valuations be based on sound and explicit economic theory and pay adequate attention to the state of ecological knowledge, including its limitations. Current economic studies have as yet been unable to achieve this goal.

ACKNOWLEDGEMENTS

Preparation of this chapter has benefited from helpful information supplied by Peter Mumby and Sabah Abdullah of the School of Biological Sciences, the University of Queensland, John Rolfe of the University of Central Queensland, Natalie Stoeckl of James Cook University, Rashid Sumaila of the University of British Columbia, and Jyothis Sathyapalan of the Centre for Economic and Social Studies, Hyderabad, as well as the editing of Karachepone Ninan. This is appreciated.

NOTES

1. The Norwegian krone estimates have not been translated into US dollars because the rate of exchange is quite variable and alters considerably over long periods.
2. Note that this is a very simplified policy perspective. It ignores the transaction costs involved in policy implementation, and fertilizer use is not the only contributor to nutrient runoff generated by the practice of agriculture.

REFERENCES

Adjeroud, M., F. Michonneau, P.J. Edmunds, Y. Chancerelle, T.L. de Loma, L. Penin, L. Thibaut, J. Vidal-Dupiol et al. (2009), 'Recurrent disturbances, recovery trajectories, and resilience of coral assemblages on a South Central Pacific reef', *Coral Reefs*, **28** (3), 775–780.

Ainsworth, C.H., J.F. Samhouri, D.S. Busch, W.W.L. Cheung, J. Dunne and T.A. Okey (2011), 'Potential impacts of climate change on Northeast Pacific marine foodwebs and fisheries', *ICES Journal of Marine Science: Journal du Conseil*, **68** (6), 1217–1229.

Allison, E.H., A.L. Perry, M.-C. Badjeck, W.N. Adger, K. Brown, D. Conway, A.S. Halls, G.M. Pilling et al. (2009), 'Vulnerability of national economies to impacts of climate change on fisheries', *Fish and Fisheries*, **10**, 173–196.

Armstrong, C.W., S. Holen, S. Navrud and I. Seifert (2012), *The Economics of Ocean Acidification – a Scoping Study*, Tromsø, Norway: Fram Centre, http://www.surfaceoa.org.uk/wp-content/uploads/2012/05/The-Economics-of-Ocean-Acidification.pdf (accessed 30 June 2015).

Brierley, A.S. and M.J. Kingsford (2009), 'Impacts of climate change on marine organisms and ecosystems', *Current Biology*, **19**, R602–R614.

Cheung, W.W.L., V.W.Y. Lam, J.L. Sarmiento, K. Kearney, R. Watson, D. Zeller and D. Pauly (2010), 'Large-scale redistribution of maximum fisheries catch potential in the global ocean under climate change', *Global Change Biology*, **15**, 24–35.

Cooley, S.R., N. Lucey, H. Kite-Powell and S.C. Doney (2012), 'Nutrition and income from molluscs today imply vulnerability to ocean acidification tomorrow', *Fish and Fisheries*, **13**, 182–215.

Costanza, R., R. de Groot, P. Sutton, S. van der Ploeg, S.J. Anderson, I. Kubiszewski, S. Farber and R.K. Turner (2014), 'Changes in the global value of ecosystem services', *Global Environmental Change*, **26**, 152–158.

Doney, S.C., V.J. Fabry, R.A. Feely and J.A. Kleypas (2009), 'Ocean acidification: The other CO_2 problem', *Annual Review of Marine Science*, **1** (1), 169–192.

Dulvy, N.K., S.I. Rogers, S. Jennings, V. Stelzenmüller, S.R. Dye and H.R. Skjoldal (2008), 'Climate change and deepening of the North Sea fish assemblage: A biotic indicator of warming seas', *Journal of Applied Ecology*, **45**, 1029–1039.

Eide, A. (2007), 'Economic impacts of global warming: The case of the Barents Sea fisheries', *Natural Resource Modeling*, **20** (2), 199–221.

Garza-Gil, M.D., J. Torralba-Cano and M. Varela-Lafuente (2011), 'Evaluating the economic effects of climate change on the European sardine fishery', *Regional Environmental Change*, **11** (1), 87–95.

Groot, R. de, L. Brander, S. van der Ploeg, R. Costanza, F. Bernard, L. Braat, M. Christie, N. Crossman et al. (2012), 'Global estimates of the value of ecosystems and their services in monetary units', *Ecosystem Services*, **1** (1), 50–61.

Hendriks, I.E., C.M. Duarte and M. Alvarez (2010), 'Vulnerability of marine biodiversity to ocean acidification: A meta-analysis', *Estuarine, Coastal and Shelf Science*, **86**, 157–164.

Hoegh-Guldberg, O., P.J. Mumby, A.J. Hooten, R.S. Steneck, P. Greenfield, E. Gomez, C.D. Harvell, P.F. Sale et al. (2007), 'Coral reefs under rapid climate change', *Science*, **318**, 1737–1742.

Kroeker, K.J., R.L. Kordas, R.N. Crim and G.G. Singh (2010), 'Meta-analysis reveals negative yet variable effects of ocean acidification on marine organisms', *Ecology Letters*, **13**, 1419–1434.

Lam, V.W.Y., W.W.L. Cheung and U.R. Sumaila (2014), 'Marine capture fisheries in the Arctic: Winners or losers under climate change and ocean acidification?', *Fish and Fisheries*, **17** (2), 335–357.

Miller, K., A. Charles, M. Berenge, K. Brander, V.F. Galluci, M.A. Gasalla, G. Munro, R. Mutugudde et al. (2010), 'Climate change, uncertainty and resilient fisheries: Institutional responses through integrative science', *Progress in Oceanography*, **87**, 338–346.

Mumby, P.J., I.A. Elliott, C.M. Eakin, W. Skirving, C.B. Paris, H.J. Edwards, S. Enríquez, R. Iglesias-Prieto et al. (2011), 'Reserve design for uncertain responses of coral reefs to climate change', *Ecology Letters*, **14** (2), 132–140.

Roff, G., S. Bejarano, Y.-M. Bozec, M. Nugues, R. Steneck and P. Mumby (2014), '*Porites* and the Phoenix effect: Unprecedented recovery after a mass coral bleaching event at Rangiroa Atoll, French Polynesia', *Marine Biology*, **161** (6), 1385–1393.

Rolfe, J. and J. Windle (2012), 'Testing benefit transfer of reef protection values between local case studies: The Great Barrier Reef in Australia', *Ecological Economics*, **81**, 60–69.

Solomon, S., D. Qin, M. Manning, Z. Chen, M. Marquis, K.B. Averyt, M. Tignor and H.L. Miller (eds) (2007), *Contribution of Working Group I to the Fourth Assessment Report of the Intergovernmental Panel on Climate Change*, Cambridge, UK: Cambridge University Press, http://www.ipcc.ch/publications_and_data/ar4/wg1/en/contents.html.

Stoeckl, N., C.C. Hicks, M. Mills, K. Fabricius, M. Esparon, F. Kroon, K. Kaur and R. Costanza (2011), 'The economic value of ecosystem services in the Great Barrier Reef: Our state of knowledge', *Annals of the New York Academy of Sciences*, **1219** (1), 113–133.

Stoeckl, N., M. Farr, D. Jarvis, S. Larson, M. Esparon, H. Sakata, T. Chaiechi, H. Lui et al. (2014), *The Great Barrier Reef World Heritage Area: Its 'Value' to Residents and Tourists*, Project 10-2 Socioeconomic Systems and Reef Resilience, Final Report to the National Environmental Research Program, Cairns, Australia: Reef and Rainforest Research Centre, http://www.nerptropical.edu.au/publication/project-102-final-report-great-barrier-reef-world-heritage-area-its-%E2%80%98value%E2%80%99-residents (accessed 23 July 2015).

Stramma, L., E.D. Prince, S. Schmidtko, J. Luo, J.P. Hoolihan, M. Visbeck, D.W.R. Wallace, P. Brandt et al. (2012), 'Expansion of oxygen minimum zones may reduce available habitat for tropical pelagic fishes', *Nature Climate Change*, **2** (1), 33–37.

Sumaila, U.R., W.W.L. Cheung, V.W.Y. Lam, D. Pauly and S. Herrick (2011), 'Climate impacts on biophysics and economics of world fisheries', *Nature Climate Change*, **1**, 449–456.

Tisdell, C.A. (2008), 'Global warming and the future of Pacific Island countries', *International Journal of Social Economics*, **35** (12), 889–903.

Tisdell, C.A. (2012), 'Economic benefits, conservation and wildlife tourism', *Acta Turistica*, **24**, 127–148.

Tisdell, C.A. (2015a), *Marine Ecosystems and Climate Change: Economic Issues*, Economics, Ecology and the Environment Working Paper No. 199, Brisbane, Australia: School of Economics, University of Queensland.

Tisdell, C.A. (2015b), *Conserving Biodiversity and Ecosystem Functions: Economic Issues*, Cheltenham, UK and Northampton, MA, USA: Edward Elgar Publishing.

9. Climate change, marine ecosystems and global fisheries

U. Rashid Sumaila, William W.L. Cheung, Philippe M. Cury and Travis Tai

CLIMATE CHANGE ALTERING MARINE ECOSYSTEM BIOPHYSICS AND ECOLOGY

The objective of this chapter is to summarize current knowledge on the potential impacts of climate change on marine ecosystems and the millions of people worldwide who depend on them. In addition, we set out to highlight the fact that different parts of the world will be impacted differently and, therefore, climate will impact people in different regions differently. Finally, we provide a number of policy recommendations to help prepare society for the changes we are already seeing and those yet to come.

Marine ecosystems are under intensifying pressures from human activities such as overfishing, pollution and habitat destruction (Pauly et al., 2002; Noone et al., 2013; Pitcher and Cheung, 2013). Climate change is an additional stress, as it is already altering the physical and chemical conditions of the ocean (Pörtner et al., 2014). Analyses of past data show that the ocean has become warmer (with a global increase in average temperature of 0.2°C between the 1950s and 1990s), has less sea ice (summer Arctic sea ice extent has been decreasing at 7.4% per decade since the 1970s), is more acidic (reduction in average pH of 0.1 unit) and has less oxygen (Gattuso et al., 2015). Warming, ocean acidification and so on are predicted to continue in the next few decades, and the rate of change further into the 21st century will depend on how carbon dioxide (CO_2) emissions are mitigated (Gattuso et al., 2015).

We have already seen biological changes in the ocean that are driven by changes in ocean conditions. Particularly, marine fishes and shellfishes that make up the bulk of the global fish catch are sensitive to temperature changes (Barange et al., 2014; Jones and Cheung, 2015). There is ample evidence showing that several species shifted their distribution, generally towards the poles and deeper waters, when the ocean warmed up (Cheung et al., 2013; Poloczanska et al., 2013). Spatial shifts in several trophic levels were observed from mesozooplankton to fish communities showing a poleward migration of sub-polar species, which are replaced by more temperate species in the North Atlantic Ocean (Beaugrand et al., 2002; Pinsky et al., 2013). Out of 664 known species of fish (of which nearly 80 are endemic), 127 alien species have become established in the Mediterranean since the beginning of the 20th century, of which 65 came through the Suez Canal (and 62 through the Strait of Gibraltar) (Coll et al., 2010).

The immigration influx of Lessepsian species (that is, marine species that migrate across the Suez Canal, usually from the Red Sea to the Mediterranean Sea) increases with warming, and the mean latitude of the distribution area of species of Atlantic origin decreases. While in the Mediterranean the endemic species distribution remained stable, that of most non-native species has spread northward by an average of 300 km since the 1980s (Coll et al., 2012). Therefore, the area of spatial overlap of the two categories of species has increased by nearly 25% in 20 years. In the North Sea, over 30 benthic macrofaunal species have been newly recorded over the last 20 years, with a distinct shift towards southern species. For instance, some bottom-dwelling fishes in the North Sea such as Atlantic cod have shifted their distributions by over 600 km towards higher latitudes and 30 m towards deeper water over the last three decades (Dulvy et al., 2008).

The timing of biological events such as spawning is shifting as well. However, the extent of such a shift may not be the same for different species, potentially leading to a mismatch between interdependent organisms such as predators and their prey. The trophodynamic of the whole ecosystem is then altered. Productivity of fish populations is also closely linked to environmental conditions. For example, the survival rate of juvenile salmon or recruitment of Atlantic cod is closely related to sea water temperature. Moreover, the distribution and amount of primary productivity, the ultimate source of food for the majority of marine life, are predicted to shift as ocean conditions change (Pörtner et al., 2014). Ocean acidification is expected to affect the growth and survival of marine organisms, particularly those that form calcium carbonate shells and skeletons, such as mussels and coral (AMAP, 2013; Gattuso et al., 2015).

NOT EVERYONE IN THE SAME BOAT

The changes described above would ultimately exact a toll on fisheries while negatively affecting the benefits to humans in many instances. Specifically, the global pattern and level of fisheries production, in terms of both weight and value, are expected to change (Sumaila et al., 2011). In January 2011, the Food and Agriculture Organization (FAO) announced that the contribution of fish to global diets had reached a new record level of about 17 kg per person per year worldwide, a rise of about 12% compared to 2000. Globally, fish provide nearly 3 billion people with at least 15% of their average per capita intake of animal protein, and support the livelihoods of millions of fishers by generating billions of dollars of wages and earnings (FAO, 2011; Teh and Sumaila, 2013). In countries such as the Seychelles, the Maldives, Kiribati, Greenland, Iceland and Namibia, fish exports are significant, in some cases constituting up to 90% of their merchandise export value. In addition, there are many other countries where up to 35% of the export values come from fish. Recreational fishing is another sector that depends on healthy oceans that would be affected by climate change. Globally, 121 million people a year participate in marine recreational activities (MRAs), generating 47 billion USD in expenditures and supporting one million jobs (Cisneros-Montemayor and Sumaila, 2010). As an example, it has been estimated that, if recreational fishing were closed completely in Florida, about 4.3 billion USD would be lost by the state's economy.

A crucial question here is: can marine fisheries continue to play their vital role as a source of food security and livelihoods in a global change context? Recent modelling studies project that we would see a large-scale redistribution of the potential fisheries catch in this century (e.g. Cheung et al., 2010; Barange et al., 2014; Jones and Cheung, 2015). Particularly, with shifts in the distribution of exploited populations and changes in primary productivity, many tropical developing countries are expected to suffer more from a reduction in the potential fisheries catch in the next few decades (Allison et al., 2009; Lam et al., 2012). Ocean acidification and de-oxygenation are likely to further reduce the overall potential catch in the future (Cheung et al., 2011; Lam et al., 2014). These changes would add to existing human pressures on the ocean, notably overexploitation, habitat loss and pollution (Noone et al., 2013). We are unlikely to see any future increases in global fish production from the current officially reported level (around 80 million tonnes per year). These would ultimately compromise the ability of the oceans to continue to provide animal protein and other ecological, economic, social and cultural values. For example, many coastal countries in West Africa are projected to suffer from losses in potential catch from climate change in addition to the losses they are already facing due to overexploitation (Lam et al., 2012). In parts of the USA, climate change may lead to reductions in the biomass of marine resources that are currently targeted by the commercial and recreational fisheries in southern states such as Louisiana and Alabama, although northern states such as Alaska may gain in catches (Ainsworth et al., 2011).

Climate change impacts on fisheries are likely to affect the most vulnerable segments of society more. For instance, commercial, subsistence and recreational fisheries are likely to be affected differently. In many countries, especially developing ones, a majority of their marine catch is landed by subsistence/artisanal fishers. In Indonesia or in Senegal, for example, 90% of the national catches of, respectively, 2.5 and 0.5 million tonnes a year are landed by small-scale fishers. In general, commercial fishers exploit resources in areas that are further from shore, while subsistence fishers usually fish closer to home. This means that the predicted impact of climate change on fish biomass distribution and migration patterns will affect subsistence and artisanal fishers more severely, with devastating consequences for the livelihoods of fishing communities around the world. Take, for instance, the approximately 200 000 people who live in the Southern Indian Ocean Biosphere Reserve, where 30% of the population rely on fish for their livelihoods, 90% of whom rely on near-shore reef-related fisheries and seaweed resources. The potential impact of ocean acidification on coral reef ecosystems on this community is likely to be huge.

As all these changes and impacts are happening, the world's population is projected to increase from the current 7 billion people to 9 billion by 2050. In addition, world incomes are expected to rise, especially in large emerging countries like China and India. The combined effects of population and income growth will be increasing demand for seafood at a time when the global marine fisheries catch has already peaked. Add to this the expected continuous increase in oil and energy prices, and the expected climate change impacts on other food production sectors such as agriculture: global cereal production is projected to suffer from slight to moderate declines and, similar to the case of fisheries, an increased disparity between the developed and developing world under climate change (Barange et al., 2014; Gattuso et al., 2015). The combined magnitude of the impacts to be expected could be truly daunting.

All of these changes are likely to affect the quality of life of many people around the world. For coastal communities, particularly in tropical developing countries, these global changes may directly reduce the availability of a major source of nutrients in their diet. In developed countries, not only may fishing communities be directly affected because of the change in resource abundance and distribution, but society in general may face more expensive seafood. In fact, given the high degree of fish exports currently, all countries would suffer the consequences of the predicted impacts of climate change on fisheries.

ARE WE PREPARED TO RESPOND TO THESE CHANGES?

Despite the great potential implications of climate change on fisheries and human well-being, current fisheries management approaches are not properly equipped to deal with the additional pressures from climate change. Fisheries policy traditionally aims for maximization of fisheries production, generally with a relatively short time horizon (usually on a year-by-year basis) and with the assumption that the characteristics of the fish stocks, the environment and regional/global socio-economic conditions will remain the same. The determination of a fishing quota based on the maximum sustainable yield target (or other variants of it) or the design and management of marine protected areas (MPAs) does not usually account for the possible changes in the biological/ecological characteristics under climate change. The ecosystem approach to fisheries (EAF) that would produce a precautionary limit reference point has not been implemented in most countries. In the case, for example, of forage fish (e.g. sardine, anchovy, sprat, herring) fisheries, which represent 30% of the world catch, the maximum sustainable yield should likely be lower by 20% to take into account food for predators (Pikitch et al., 2014). This precautionary approach would serve to buffer, in many respects, negative climatic impacts. Moreover, current management does not account for the potential effects of different scenarios of long-term social and economic changes. This may compromise the long-term effectiveness of existing fisheries management and marine conservation measures. The problem is even more acute in areas where fisheries and ocean management is poor or virtually non-existent while fish stocks and marine ecosystems are under great pressures from human impacts such as overexploitation, habitat loss and pollution (Cheung and Sumaila, 2015). Unfortunately, fisheries in these areas are possibly amongst the ones to be most strongly affected by climate change.

The need to develop effective mitigation and adaptation strategies and policies to deal with the impending impacts of climate change cannot be overemphasized. These policies will require credible scientific assessments of the close linkages between the physical, ecological and socio-economic systems that fisheries depend on. We will need comprehensive understandings of such linkages at all levels to decide what should be done, in both the short and the long term. However, the lack of full knowledge should not be an excuse to delay policy actions. Instead, we should develop policies that are robust to the uncertainties, with careful consideration of costs and risk (Polasky et al., 2011; Cheung et al., 2016a). Meanwhile, the Intergovernmental Science-Policy Platform on Biodiversity and Ecosystem

Services (IPBES) will also contribute to the building of knowledge and capacity for biodiversity and ecosystem services to develop adequate and articulated responses to these complex issues (Díaz et al., 2015).

POLICY RECOMMENDATIONS FOR CHANGE

In order to provide meaningful recommendations, it is important to understand the key characteristics of climate change and its potential impacts on fisheries. First, climate change is a 'big' commons problem, where what we do individually, usually without considering the effects on other people, actually affects all citizens of the world. It therefore suffers hugely from what economists denote as externality, that is, a situation in which the action of one party has a consequence on other people, and the party taking the actions does not take into account the effects of its actions on others. Second, for most people, climate change is difficult to grasp and deal with. The issues get more complicated to communicate to the public, with the direct and indirect effects of greenhouse gases emissions on biogeochemistry, biology, ecology and fisheries in the ocean. Third, the potential impacts of climate change are perceived to be too uncertain for many people. Fourth, the most severe consequences of climate change will be felt in the distant future, thereby affecting future generations, who currently do not have a vote, more negatively. Fortunately, many in the current generation care about the well-being of future generations, which is all that is needed for current policies to consider the interests of future generations (Ekeland et al., 2015).

The above characteristics imply that developing an effective climate change response is challenging, and requires that all segments of society – individuals, civil society, companies, non-governmental organizations, governments – change their behaviour with respect to the use of fossil fuels and energy. It means that we need to develop institutions at all levels of society, in particular at the global level, to help mitigate and adapt various economic sectors, including fisheries, to climate change. Consequently, our recommendations are targeted to specific groups and levels of society. First, we call for the need to take a long-term comprehensive view of the issues in developing marine ecosystem management policies. This includes assessing major management and conservation policies on their adaptability under a changing climate (Sumaila and Cheung, 2010). This can be done through incorporating scenario analysis in addition to routine fisheries assessment (e.g. Mullon et al., 2016). We should prioritize climate change mitigation and adaptation policies, particularly those that generate 'co-benefits', for example improving fisheries and ocean management, and building adaptive capacity for local communities. Second, we need to

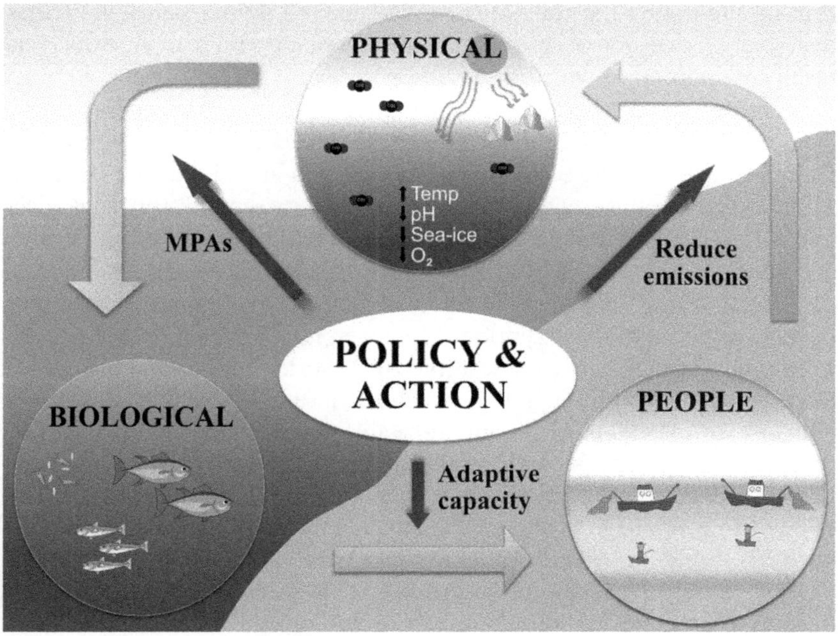

Note: With scenario development, we can use this framework to develop models and predict future conditions to determine pathways that will benefit the environment and society for generations to come.

Figure 9.1 Adaptation to and mitigation of climate change and ocean acidification impacts on marine ecosystems and global fisheries can be implemented on various fronts using policy and action from all segments of society, that is, individuals, civil society, companies, NGOs and governments

do better in communicating to the public about the potential impacts of climate change on fisheries and our well-being. Third, we should develop mechanisms that fairly spread the costs and benefits among different countries and different generations. Finally, we need to increase the adaptive capacity of our social, ecological, economic and legal systems to climate change (Figure 9.1).

During the last few decades, marine ecosystems have been changing at surprising rates in a global change context. Under strong and increasing anthropogenic pressures, the complexity of dynamic interactions (e.g. between overexploitation, climate change, species migrations and market globalization) questions our ability to efficiently understand, predict and manage marine resources in a changing context. Building scenarios can help in integrating the state-of-the-art science to deliver plausible stories on ecosystem responses to global change, which would in turn help us reach common objectives.

While we cannot perfectly predict the future, we can better prepare for it by defining desired objectives with a recognition of this limitation. Defining and targeting ecosystem objectives that are relevant for the EAF would accelerate the exchange of information between physical, ecological and social scientists and help to define robust pathways, resulting in better decisions (Moss et al., 2010). In a context where the demand to address complex issues is considerable, the interdisciplinary approach becomes the rule rather than the exception. Moreover, it would help in meeting the commitments on fisheries made at the 2002 World Summit on Sustainable Development. By bringing together realistic environmental constraints with desired societal goals, we might be able to reach ecosystem objectives that have been shown recently to be impossible to achieve (e.g. to significantly reduce the rate of biodiversity loss by 2010: Butchart et al., 2010) or difficult to reach (e.g. to restore fish stocks to maximum sustainable yield by 2015). Building objective-oriented scenarios could make a more effective use of science and information in policy development and decision making while achieving internationally agreed objectives for the EAF.

Building scenarios requires the involvement of a large scientific community that will discuss and adopt a global scientific strategy. There is also the need to capitalize on, integrate and distribute available information on marine ecosystems in a more efficient and convincing manner. Knowledge-based centres that bring together all the information to support decision making on ecosystems seem to be a prerequisite for avoiding fragmented views and poor data integration (e.g. using an ecoscope: Cury, 2004). The challenge is to make most ecological data freely available, as is the case with most physical data.

Scientists will have to play a more visible role in bringing their work to public attention and into policy. Science is at the core of this new policy, not just to inform on what is probable but also to envision future pathways and account for critical uncertainties. This new role will require the improvement of scientific communication using scenarios. Scientists have been communicating through target or limit reference points (e.g. maximum sustainable yield or threshold limit fish biomass) using maps, texts and figures. Communicating with scenarios constitutes a new challenge, as a daunting amount of information needs to be conveyed to a large audience in a very limited time within the framework of the ecosystem approach to fisheries. A new communication strategy is required if we want to convey insights from scenario analysis to stakeholders whose role appears to be central when building socio-ecological scenarios, especially for management issues at a local scale. In such cases, scenarios are often based on socio-economic assumptions and policy options knowing ecological dynamics, which are then used to study the consequences of alternative management policies on ecosystems. For this purpose, we will need dedicated scenario laboratories, that is, places where we can visualize, experiment and discuss facts, assumptions and options among stakeholders, and envision the future using scenarios.

More specifically: (1) governments need an adaptive approach to performing large-scale experiments on fisheries and marine resource exploitation to understand how to deal with surprises, which will always be part of ecosystem changes; (2) governments need to develop institutions and incentive structures that will help ensure cooperation and coordination within and between countries on the use of marine resources, which are by definition shared common property resources (Miller et al., 2013; Sumaila, 2013), and it will be necessary to involve stakeholders and communities on policy formulation and implementation; (3) the multifaceted and cross-cutting nature of climate change impacts and the responses needed requires that government ministries coordinate their policies and actions, and collaborate in ways that they never used to in the past. This requires governments around the world to remove institutional barriers to collaboration between government departments and to develop scenario building as national and global frameworks for managing marine resources.

The mitigation and adaptation of fisheries to climate change will require investments, both public and private. Hence, governments should consider putting in place a climate change fund that would mobilize resources from all possible sources for investment in these important activities. Such investments are needed to organize skill training programmes for fishers in order to allow them to diversify their livelihood possibilities and thereby make it easier for them to switch among economic activities. Similarly, there is a strong need for knowledge transfer and capacity building, especially for highly vulnerable developing countries. Research is needed on: mitigation strategies; ecosystem-based fisheries management; and climate change impacts on inland fisheries and aquaculture. It would be prudent for governments to identify and implement 'no-regret' policy measures immediately, for example policies that rebuild depleted ecosystems and improve habitat quality.

Scientific work is mainly directed to the publication of articles in journals. With humanity facing bewildering and worrying changes that are increasingly challenging human ingenuity, a new ethics for scientists is emerging that will change the way to envision a scientific career in a global change context. Scientists must remain independent and communicate their results to the public and policy makers. They must convey, in more efficient ways, 'inconvenient truths' and 'successful management stories' to the public. In this way, science will play an even more important role in detecting, understanding and quantifying those changes and communicating them to a larger audience. Collaborations between society and scientists could facilitate such communications. Scientists should also be able to address the proper scale at which global change may be felt and how this may be addressed. The limitations of the current literature lie mostly in terms of the appropriateness of scale and the proper context within which to understand the appropriate scales at which to implement policies while all major sectors of the global economy are facing pressures and stresses from climate change all at the same time.

The track record of civil society (NGOs, the press, private foundations, etc.) in dealing with societal problems such as poverty reduction, child abuse, and tackling environmental degradation means that there is a big role for civil society in dealing with the current and predicted challenges to fisheries in the face of climate change. They can contribute by organizing campaigns to educate the public on the effects of climate change on fisheries and help inform the political debate. Similarly, fishing, processing and retail companies active in the fishing sector need to develop adaptation plans that will help them weather the storm when it comes to climate change and its impact on global fisheries.

It is a well-known fact that, in general, individuals, groups or society at large often take a short-term and narrow view in deciding actions or developing policies, which does not match both the temporal and the spatial scales of climate change, for example deciding next year's catch quotas rather than thinking of flows of catch quotas through time. This probably leads to policies that are maladapted to climate change.

Ultimately, to build climate resilient marine ecosystems and global fisheries, we need to transform their management by increasing incentives for community engagement (Cinner et al., 2016), deploying marine protected areas (Sumaila et al., 2015; Cheung et al., 2016b), and putting in place only sustainability enhancing public policies by avoiding harmful ones such as the provision of capacity enhancing fisheries subsidies (Sumaila et al., 2016).

ACKNOWLEDGEMENTS

The authors would like to thank Daniel Pauly, Gabriel Reygondeau, Jorge Sarmiento and Vicky Lam for discussions. U. Rashid Sumaila and William W.L. Cheung want to thank the OceanCanada Partnership, and William W.L. Cheung the Nereus project for support.

REFERENCES

Ainsworth, C.H., J.F. Samhouri, D.S. Busch, W.W.L. Cheung, J. Dunne and T.A. Okey (2011), 'Potential impacts of climate change on Northeast Pacific marine foodwebs and fisheries', *ICES Journal of Marine Science*, **68**, 1217–1229.

Allison, E.H., A.L. Perry, M.-C. Badjeck, W.N. Adger, K. Brown, D. Conway, A.S. Halls, G.M. Pilling et al. (2009), 'Vulnerability of national economies to the impacts of climate change on fisheries', *Fish and Fisheries*, **10**, 173–196.

AMAP (Arctic Monitoring and Assessment Programme) (2013), *Assessment 2013: Arctic Ocean Acidification*, Oslo: AMAP.

Barange, M., G. Merino, J.L. Blanchard, J. Scholtens, J. Harle, E.H. Allison, J.I. Allen, J. Holt et al. (2014), 'Impacts of climate change on marine ecosystem production in societies dependent on fisheries', *Nature Climate Change*, **4**, 211–216.

Beaugrand, G., P.C. Reid, F. Ibañez, J.A. Lindley and M. Edwards (2002), 'Reorganization of North Atlantic marine copepod biodiversity and climate', *Science*, **296**, 1692–1694.

Butchart, S.H.M., M. Walpole, B. Collen, A. van Strien, J.P.W. Scharlemann, R.E.A. Almond, J.E.M. Baillie, B. Bomhard et al. (2010), 'Global biodiversity: Indicators of recent declines', *Science*, **328** (5982), 1164–1168.

Cheung, W.W.L. and U.R. Sumaila (2015), 'Economic incentives and overfishing: A bioeconomic vulnerability index', *Marine Ecology Progress Series*, **530**, 223–232.

Cheung, W.W.L., V.W.Y. Lam, J.L. Sarmiento, K. Kearney, R. Watson, D. Zeller and D. Pauly (2010), 'Large-scale redistribution of maximum fisheries catch potential in the global ocean under climate change', *Global Change Biology*, **16**, 24–35.

Cheung, W.W.L., J. Dunne, J.L. Sarmiento and D. Pauly (2011), 'Integrating ecophysiology and plankton dynamics into projected maximum fisheries catch potential under climate change in the Northeast Atlantic', *ICES Journal of Marine Science*, **68**, 1008–1018.

Cheung, W.W.L., R. Watson and D. Pauly (2013), 'Signature of ocean warming on global marine fisheries', *Nature*, **497** (7449), 365–368.

Cheung, W.W.L., T.L. Frolicher, R.G. Asch, M.C. Jones, M.L. Pinsky, G. Reygondeau, K.B. Rodgers, R.R. Rykaczewski et al. (2016a), 'Building confidence in projections of the responses of living marine resources to climate change', *ICES Journal of Marine Science*, **73** (5), 1283–1296.

Cheung, W.W.L., M.C. Jones, V.M.Y. Lam, D.D. Miller, Y. Ota, L. Teh, Ussif and R. Sumaila (2016b), 'Transform high seas management to build climate-resilience in marine seafood capacity', *Fish and Fisheries*, DOI: 10.1111/faf.12177.

Cinner, J.E., C. Huchery, M.A. MacNeil, N.A.J. Graham, T.R. McClanahan, J. Maina, E. Maire, J.N. Kittinger et al. (2016), 'Bright spots among the world's coral reefs', *Nature*, **535**, 416–419.

Cisneros-Montemayor, A.M. and U.R. Sumaila (2010), 'A global valuation of ecosystem-based marine recreation', *Journal of Bioeconomics*, **12**, 245–268.

Coll, M., C. Piroddi, J. Steenbeek, K. Kaschner, F. Ben Rais Lasram, J. Aguzzi, E. Ballesteros, C.N. Bianchi et al. (2010), 'The biodiversity of the Mediterranean Sea: Estimates, patterns, and threats', *PLoS ONE*, **5** (8), e11842.

Coll, M., C. Piroddi, C. Albouy, F. Ben Rais Lasram, W.W.L. Cheung, V. Christensen, V.S. Karpouzi, F. Guilhaumon et al. (2012), 'The Mediterranean Sea under siege:

Spatial overlap between marine biodiversity, cumulative threats and marine reserves', *Global Ecology and Biogeography*, **21**, 465–480.

Cury, P. (2004), 'Tuning the ecoscope for the ecosystem approach to fisheries', *Marine Ecology Progress Series*, **274**, 272–275.

Díaz, Sandra, Sebsebe Demissew, Julia Carabias, Carlos Joly, Mark Lonsdale, Neville Ash, Anne Larigauderie, J.R. Adhikari et al. (2015), 'The IPBES conceptual framework – connecting nature and people', *Current Opinion in Environmental Sustainability*, **14**, 1–16.

Dulvy, N.K., S.I. Rogers, S. Jennings, V. Stelzenmüller, S.R. Dye and H.R. Skjoldal (2008), 'Climate change and deepening of the North Sea fish assemblage: A biotic indicator of warming seas', *Journal of Applied Ecology*, **45**, 1029–1039.

Ekeland, I., L. Karp and U.R. Sumaila (2015), 'Equilibrium resource management with altruistic overlapping generations', *Journal of Environmental Economics and Management*, **70**, 1–16.

FAO (Food and Agriculture Organization of the United Nations) (2011), *The State of World Fisheries and Aquaculture 2010*, Rome: FAO.

Gattuso, J.-P., A. Magnan, R. Billé, W.W.L. Cheung, E.L. Howes, F. Joos, D. Allemand, L. Bopp et al. (2015), 'Contrasting futures for ocean and society from different CO_2 emissions scenarios', *Science*, **349** (6243).

Jones, M.C. and W.W.L. Cheung (2015), 'Multi-model ensemble projections of climate change effects on global marine biodiversity', *ICES Journal of Marine Science*, **72**, 741–752.

Lam, V.W.Y., W.L. Cheung, W. Swartz and U.R. Sumaila (2012), 'Climate change impacts on fisheries in West Africa: Implications for economic, food and nutritional security', *African Journal of Marine Science*, **34**, 103–117.

Lam, V.W., W.W.L. Cheung and U.R. Sumaila (2014), 'Marine capture fisheries in the Arctic: Winners or losers under climate change and ocean acidification?', *Fish and Fisheries*, DOI: 10.1111/faf.12106.

Miller, K.A., G.R. Munro, U.R. Sumaila and W.W.L. Cheung (2013), 'Governing marine fisheries in a changing climate: A game-theoretic perspective', *Canadian Journal of Agricultural Economics*, **61** (2), 309–334.

Moss, R.H., J.A. Edmonds, K.A. Hibbard, M.R. Manning, S.K. Rose, D.P. van Vuuren, T.R. Carter, S. Emori et al. (2010), 'The next generation of scenarios for climate change research and assessment', *Nature*, **463** (11), 746–756.

Mullon, C., F. Steinmetz, G. Merino, J.A. Fernandes, W.W.L. Cheung, M. Butenschön and M. Barange (2016), 'Quantitative pathways for Northeast Atlantic fisheries based on climate, ecological-economic and governance modelling scenarios', *Ecological Modelling*, **320**, 273–291.

Noone, K.J., U.R. Sumaila and R.J. Diaz (eds) (2013), *Managing Ocean Environments in a Changing Climate: Sustainability and Economic Perspectives*, London: Elsevier.

Pauly, D., V. Christensen, S. Guénette, T.J. Pitcher, U.R. Sumaila, C.J. Walters, R. Watson and D. Zeller (2002), 'Towards sustainability in world fisheries', *Nature*, **418**, 685–695.

Pikitch, E.K., Konstantine J. Rountos, Timothy E. Essington, Christine Santora, Daniel Pauly, Reg Watson, U. Rashid Sumaila, P. Dee Boersma et al. (2014), 'The global contribution of forage fish to marine fisheries and ecosystems', *Fish and Fisheries*, **15** (1), 43–64.

Pinsky, M.L., B. Worm, M.J. Fogarty, J.L. Sarmiento and S.A. Levin (2013), 'Marine taxa track local climate velocities', *Science*, **341**, 1239–1242.

Pitcher, T.J. and W.W.L. Cheung (2013), 'Fisheries: Hope or despair?', *Marine Pollution Bulletin*, **74**, 506–516.
Polasky, S., S.R. Carpenter, C. Folke and B. Keeler (2011), 'Decision-making under great uncertainty: Environmental management in an era of global change', *Trends in Ecology and Evolution*, **26**, 398–404.
Poloczanska, E.S., C.J. Brown, W.J. Sydeman, W. Kiessling, D.S. Schoeman, P.J. Moore, K. Brander, J.F. Bruno et al. (2013), 'Global imprint of climate change on marine life', *Nature Climate Change*, **3**, 919–925.
Pörtner, H.O., D. Karl, P.W. Boyd, W.W.L. Cheung, S.E. Lluch-Cota, Y. Nojiri, D. Schmidt, P.O. Zavialov et al. (2014), 'Ocean systems', in C.B. Field, V.R. Barros, D.J. Dokken, K.J. Mach, M.D. Mastrandrea, T.E. Bilir, M. Chatterjee, K.L. Ebi et al. (eds), *Climate Change 2014: Impacts, Adaptation, and Vulnerability, Part A: Global and Sectoral Aspects*, New York: Cambridge University Press, pp. 411–484.
Sumaila, U.R. (2013), *Game Theory and Fisheries: Essays on the Tragedy of Free for All Fishing*, London: Routledge.
Sumaila, U.R. and W.W.L. Cheung (2010), *Cost of Adapting Fisheries to Climate Change*, World Bank Discussion Paper No. 5, Washington, DC: International Bank for Reconstruction and Development/World Bank.
Sumaila, U.R., W.W. Cheung, V.W. Lam, D. Pauly and S. Herrick (2011), 'Climate change impacts on the biophysics and economics of world fisheries', *Nature Climate Change*, **1**, 449–456.
Sumaila, U.R., Vicky W.Y. Lam, Dana D. Miller, Louise Teh, Reg A. Watson, Dirk Zeller, William W.L. Cheung, Isabelle M. Côté et al. (2015), 'Winners and losers in a world where the high seas is closed to fishing', *Scientific Reports*, **5**, article 8481.
Sumaila, U.R., V. Lam, F. Le Manach, W. Swartz and D. Pauly (2016), 'Global fisheries subsidies: An updated estimate', *Marine Policy*, **69**, 189–193.
Teh, L.C. and U.R. Sumaila (2013), 'Contribution of marine fisheries to worldwide employment', *Fish and Fisheries*, **14** (1), 77–88.

10. Extreme drought and California's water economy: challenges and opportunities for building resilience

Kathleen A. Miller

INTRODUCTION

Water resource management in the U.S. State of California has long been a story about struggles to build resilience while serving many competing demands in a highly variable climate. Droughts and floods have posed recurring challenges for the state's citizens, and efforts to address those risks played a dominant role in driving development of the state's water management infrastructure and supporting institutional arrangements. In many respects these efforts have built considerable resilience to climate variability by reducing flood risks, allowing water to be stored for later use, and moving it around the state to augment supplies in water-scarce areas. There are weaknesses in that resilience, however, that have left some communities and water-related values quite vulnerable to protracted severe droughts, as occurred in 1987–1992 and 2012–2015. Such events, in turn, have spurred new legislation and proposals for institutional innovations that will be important for ameliorating future vulnerabilities to droughts and to the impacts of climate change on the state's water resources. In particular, the most recent drought has spurred important changes in the state's approach to groundwater management, better documentation of surface water withdrawals, and numerous local initiatives to more sustainably manage available water supplies.

Although the state and federal governments play large roles in management of California's water resources, the state's water infrastructure and institutions grew up over time in a piecemeal fashion to solve specific problems and to forestall conflicts that might arise between rival water uses or incompatible modifications of a river system. The result of this process has been a complicated multi-layered water management system involving a mixture of public entities ranging from municipal and agricultural water providers to agencies of the state and federal governments, as well as a large number of private owners of water rights that were established at different points in time under different sets of rules. California recognizes three broad classes of surface water rights (Gray et al., 2015):

- Riparian rights can be held by owners of land directly adjacent to a natural water source. They are the most senior rights; they are not defined as to quantity of use, and can be regulated by the state only to prevent unreasonable or wasteful use.
- Pre-1914 appropriative rights were claimed under the principle of *first in time, first in right*, but before enactment of the modern water code in 1914. Documentation for these rights varies in quality, consisting primarily of court records or claims filed in local jurisdictions (Hanemann et al., 2015).
- Post-1914 appropriative rights were established under the state's licensing system, and are administered under the authority of the State Water Resources Control Board (SWRCB).

Groundwater rights have remained largely unquantified and unregulated except in basins where a court proceeding has adjudicated the pumping rights of overlying landowners. Such diversity in the nature of water rights, together with differences in physical access to water and the means of controlling it, leaves some water users facing greater vulnerability to droughts than others. In addition, throughout the 19th and early 20th centuries, California water policy focused on securing benefits for miners, farmers and cities. Little thought was given to the protection of natural aquatic environments and the ecological, recreational and flood mitigation services they could support. Perspectives have shifted dramatically in recent decades. Better scientific understanding of these natural systems and growing public desire for their protection have led to environmental regulations and fostered citizen support for effective representation of environmental values in public water planning forums. However, water rights for instream flows have been established only recently, and environmental water allocations may be trumped by more senior diversion rights during droughts.

Challenges now facing the state's water policy community include the need to balance the goals of: providing reliable, affordable and high-quality water to a rapidly growing population; protecting that population from loss of life and property during floods; restoring aquatic ecosystems long damaged by blocked and dewatered streams; and preserving the open landscapes, free-flowing river reaches, and productive farms that are valued by the state's citizens and visitors alike. The state's recent experiences with severe droughts demonstrate how earlier decisions on land use, infrastructure investment and water law shape both vulnerabilities to extreme climate events and options for building resilience to anticipated climate change. These experiences also shed light on the meaning of resilience and the processes by which water management regimes evolve in response to multiple stresses.

A snapshot of water use practices in the state will help readers to understand the challenges and opportunities involved in building a resilient water economy. California leads the United States on many measures of economic activity. It is the nation's most populous state, and has the largest state gross domestic product and the most valuable agricultural sector. Management of water resources plays an important and highly visible role in supporting that economic activity. In particular, water is critical for California's agricultural sector, which produces 11% of the value of U.S. agricultural output as measured by farm-level sales (USDA-ERS, 2015). The state's irrigated farms produce half of the fruit, nuts and fresh vegetables consumed in the nation. Together with milk production, which also relies on irrigation, these high-value crops contributed more than 70% of the approximately US$43 billion value of the agricultural products sold in California in 2012 (USDA, 2012).

Because very little rain falls during the growing season, California's agriculture depends heavily on irrigation water provided primarily by surface water projects which collect, transport and distribute water from rivers fed by runoff from the state's mountain ranges. Groundwater pumping supplements those supplies especially during drought years, and in some areas groundwater depletion accelerated markedly during the most recent drought. Agriculture dominates California's water economy, accounting for approximately 80% of the state's consumption of water by human activities (Mount et al., 2014).[1]

Over the course of the past century and a half, entities ranging from small groups of farmers to agencies of the state and federal government built numerous water projects to provide secure water supplies and reduce potential flood damages. These efforts have transformed California's river basins into highly engineered systems supporting irrigated agriculture and urban settlement. California is now home to an elaborate array of water storage, conveyance and distribution projects. These systems deliver irrigation water to approximately 9 million acres (3.64 million hectares) of highly productive farmland and move vast quantities of water from areas with abundant water supplies to relatively dry and heavily populated parts of the state.

Two such projects, the federal Central Valley Project (CVP) and the State Water Project (SWP), annually transport an average of approximately 6 km^3 of freshwater from northern California to farms and cities in southern and central California (Healey et al., 2008). The State Water Project is especially important for augmenting municipal water supplies for the heavily urbanized Southern California Coastal Plain, which is home to approximately half of California's 39 million residents. On average, SWP deliveries account for about 30% of the water supplies for urban southern California, with the Metropolitan Water District (MWD) of Southern California being the largest contractor for SWP deliveries (MWD, n.d.). Those deliveries are highly variable from year to year, however, and there have been deep cuts in allocations during the most severe drought years (CDWR, 2015; State of California, 2015).

Diversions from the Colorado River into southern California bring another 5.4 km^3 of water into that part of the state.[2] California also has an extensive system of levees and floodways designed to shunt floodwaters away from settlements, but there are serious questions about the vulnerability of those levees to large floods, earthquakes and rising sea levels (Healey et al., 2008; Jones, 2016; Zellmer and Klein, 2016).

The Central Valley, with approximately 4 million irrigated acres (1.6 million hectares), is a primary hub for the state's irrigated agriculture, which dominates the Valley's landscape. Irrigation water is provided by a mixture of large and small surface water projects, as well as by groundwater pumping from the Valley's extensive aquifers. The Central Valley comprises two major drainage basins. The Sacramento River flows southward to drain the northern part of the Valley, while the smaller San Joaquin River flows northward. The two rivers meet in the Sacramento–San Joaquin Delta, which drains into the San Francisco Bay.

The Sacramento Basin contributes roughly three-quarters of the discharge from the Delta, but occupies only about one-third of the Central Valley's land area. Both rivers draw a substantial portion of their volume from runoff from the surrounding mountains. Peak streamflows, fed by melting mountain snowpacks, occur in the spring and early summer.

The Delta is a complex tangle of meandering waterways and small islands protected by levees, many of which have sunk below sea level as their rich organic soils have been depleted by farming and are no longer being replenished by regular inundation. The intakes for both the Central Valley Project and the State Water Project are located in the Delta. This has long been recognized as a source of vulnerability for the state's water supply system. According to Richard Norgaard,

> Water engineers have always seen the Delta as an obstacle, starting with the first advocates of grand plans to bring water to southern California in the nineteenth century. Encouraging water to flow from north to diversion pumps in the south of the Delta, through a maze of islands through which water naturally flows west, was long recognized as a bad idea. (Norgaard, 2016: 226)

The current configuration of the Delta water export infrastructure requires substantial water releases from reservoirs in the Sacramento Basin to prevent salty water from backing up through the estuary to the pump intakes. This arrangement is vulnerable to low flows during extended droughts and, increasingly, to sea-level rise caused by global climate change. In addition, earthquakes, floods or poor maintenance could cause failure of the aging levees that protect islands in the Delta. "Major failure among the Delta levees could draw a massive influx of sea water from the San Francisco Bay into the freshwater parts of the Delta, which could render it unusable as a central link in the State's major freshwater-conveyance systems" (Dettinger et al., 2015: 2087). Operation of the pumps also has proved highly detrimental to fish species migrating through the Delta, such as the juvenile salmon that must navigate through the Delta to reach the ocean where they can grow and mature. In 1993 and 1994, in the wake of the 1987–1992 drought, two fish species in the Delta were listed as requiring protection under the Endangered Species Act (ESA) (Water Education Foundation, n.d.). That federal action required more stringent restrictions on pumping operations and set in motion a long and contentious planning process focused on addressing the Delta's environmental problems (Norgaard, 2016).

There have been several proposals for bypassing the Delta with a peripheral canal or set of tunnels, which would place the withdrawal point farther upstream and avoid problems with saline intrusion. But such proposals would be very costly and would have other adverse consequences including reduced water quality for farms in the Delta and impacts on the Delta's ecology. Political standoffs and conflicting objectives continue to hamper resolution of the Delta's vulnerabilities.

Despite these very real challenges and vulnerabilities, there have been a number of positive developments in recent years that will promote greater water supply sustainability. These include extensive programs by southern California's urban water suppliers to reduce their dependence on water from the Delta by reducing demand growth and increasing local water storage, including recharging groundwater aquifers with wet-year surplus supplies, storm runoff and recycled wastewater.

CALIFORNIA DROUGHTS IN PERSPECTIVE

Severe droughts have repeatedly tested the resilience of California's water management system. Most recently, the effects of an epic drought in 2012–2015 are still lingering as this chapter is being written. Some notable earlier droughts include: water-year 1924 (October 1, 1923 – September 30,

1924), which holds the record for lowest statewide precipitation (California Climate Tracker, 2016); the six-year drought of 1929–1934; the sharp two-year drought of 1976–1977, which included record-low statewide runoff in 1977; and the six-year drought period of 1987–1992 (Jones, 2015). The historical record offers only a limited picture of the natural processes that determine the frequency and intensity of California droughts. Reconstructions of paleoclimate conditions from tree rings and geologic evidence provide a record of numerous droughts of varying durations and intensity, including some considerably longer than have been documented in the instrumental record (Lindström, 1990; Meko et al., 2014; Jones, 2015).

The early 20th century droughts played a role in fueling public support for major water supply projects, including the Central Valley Project, which was built in stages starting in the 1930s by the U.S. Bureau of Reclamation (Reclamation, 2015a). The 1987–1992 drought, on the other hand, ushered in a period of institutional change that set the stage for responses to the most recent drought. Responses to the 2012–2015 drought have continued the pattern of taking advantage of a current crisis to build resilience to future climatic extremes. But there are new considerations now. These include increasingly constrained water development options, the growing need to balance out-of-stream water uses against the values provided by healthy aquatic environments, and a growing recognition that climate change may make droughts, floods and years with paltry snowpacks a more common occurrence in the future.

California's climate is highly variable across both time and space. It is characterized by strongly seasonal precipitation with cool wet winters and hot dry summers, high inter-annual variability in total precipitation, and striking contrasts across the state in local hydroclimatic regimes. Most of the state's water is supplied by a relatively small number of large cool-season storms (typically between November and March), while very little rain falls during the summer growing season. As measured by the coefficient of variation of annual precipitation, California has the most variable hydroclimate of any part of the United States (Dettinger et al., 2011). Droughts and years with heavy runoff and potential flooding recur frequently.

Spatial variability in precipitation is driven largely by the state's topography. Mountain ranges along the coast and the state's eastern flank surround the low-lying and nearly flat 58000 km^2 Central Valley, which occupies roughly 14% of the area of the state (Figure 10.1). It has been said that "Orography is destiny with respect to the state's precipitation" (Jones, 2016, p. 208). Indeed annual precipitation averages about 1000 to 2000 mm/yr over most of the northern coastal ranges and along the spine of the Sierra Nevada and Cascade Mountains, with much higher amounts

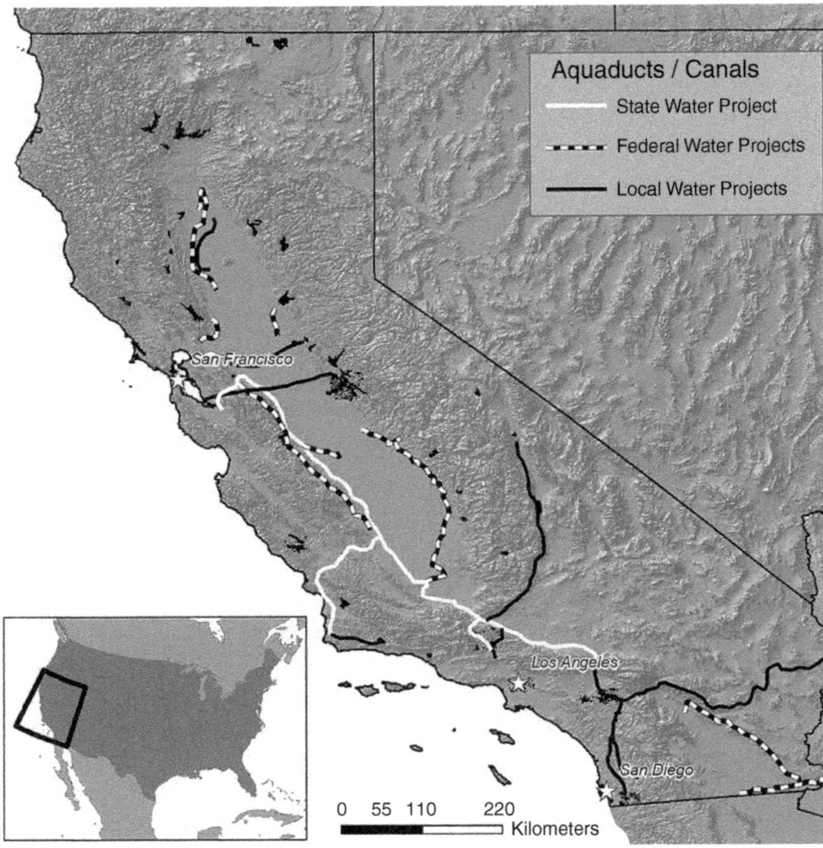

Source: Map created by Jennifer Boehnert, National Center for Atmospheric Research.

Figure 10.1 California topography and major water storage and conveyance projects

occurring in some high-elevation locations. The Central Valley, in contrast, receives roughly 500–750 mm/yr at its far northern edge and only in the neighborhood of 125–250 mm/yr throughout the southern two-thirds of the Valley. Elsewhere in southern California, there are large desert expanses receiving less than 125 mm/yr (Western Regional Climate Center, 2015).

Historically, most of California's water supply projects were designed to work in concert with the regular seasonal accumulation and melting of the state's mountain snowpacks. In essence, the snowpacks represent key water storage reservoirs – holding flows back in the winter when they might otherwise pose a flood risk and the water is not wanted for human uses, and releasing it to flow into the valleys just at the start of the irrigation season.

The warming anticipated from global climate change is likely to disrupt this long-established pattern. Warmer temperatures will inevitably alter the dynamics of snow accumulation and melting, leading to impairment of the natural water storage capacity of high-elevation catchments and alteration of seasonal river flow regimes (Adam et al., 2009; Stewart, 2009). The loss of snowpack storage could spell reduced late summer streamflows, reduced water availability for irrigation and other human uses, and higher water temperatures. Already, the relative contribution of snowmelt to annual flows has declined somewhat over the past century. Runoff seasonality is quite pronounced in the San Joaquin Basin, where April to July runoff accounted for roughly 65–70% of the annual total, on average, over the past century. The April to July runoff proportion in the Sacramento Basin has been lower, roughly in the neighborhood of 35–40%, because the lower elevation mountains there allow a larger fraction of runoff to occur in the winter period. Both basins have seen a trend toward earlier runoff over the past century (Roos and Sahota, 2012).

Shorter snow seasons also are likely to lead to reduced summer soil moisture in mountainous areas and increased vegetation stress. That would contribute to an increased likelihood of wildfires and forest disturbance due to insect outbreaks and drought-induced mortality (e.g. Allen et al., 2010). Such disturbances would further affect watershed hydrology, potentially leading to flashier runoff, higher stream temperatures, heavier sediment loads and associated water quality degradation (Adams et al., 2012; Luce et al., 2012).

In 2015, California endured the fourth consecutive year of a severe drought. Not only had the period between fall 2011 and fall 2015 been the driest consecutive 48 months since recordkeeping began in 1895, but it was also the warmest 48-month period as measured by statewide average temperatures (NOAA National Centers for Environmental Information, 2015).

In particular, record warm temperatures in 2014 and 2015 compounded the effects of scanty precipitation to deprive the state of its most important natural reservoir: water stored in the winter snowpack that normally accumulates in the Sierra Nevada and Cascade Mountains. As of April 1, 2015 the measured statewide snowpack stood at only 5% of the long-term average for that date (CDWR, 2015). A handful of heavy rainstorms had left some of the state's artificial surface reservoirs in better condition, but all major reservoirs held less water than normal for that time of year. Without a normal snowpack there was little prospect of refill. Although statewide precipitation recovered to 77% of normal in water-year 2015

(October 1, 2014 – September 30, 2015), the impacts of the extended drought and high temperatures on soil moisture and evapotranspiration led to the second-lowest estimated statewide annual runoff total since the beginning of the 20th century (USGS, 2015). By November 2015, the U.S. Drought Monitor reported that more than two-thirds of California had endured extreme to exceptional drought conditions for at least 22 months (U.S. Drought Monitor, 2015).

This severe drought episode has produced very uneven impacts on different water users and sections of the state as well as on natural ecosystems versus managed landscapes. These differences can be traced to the state's complex geography, the configuration of its water storage and delivery infrastructure, and its imperfectly administered mixture of prior appropriation and riparian surface water rights (Hanemann et al., 2015). Other factors include the state's limited regulation of groundwater withdrawals and historically spotty protection for aquatic ecosystems. Efforts to mitigate the impacts of the drought have included emergency regulations and a variety of voluntary water conservation activities. California's recent severe drought years have worsened the ongoing challenges faced by the state's water management community, but the experience also has spurred new investments, changes in water management practices and institutional innovations that are gradually steering the state towards greater resilience to future climatic stresses.

THE EVOLVING MANAGEMENT CONTEXT

The 1987–1992 drought can be seen as a turning point in California's water policy environment. Prior to that event, the development of new water supply projects had generally forestalled any serious water shortages for the state's major cities and centers of agricultural production. As the severe multi-year drought progressed, there were mounting fears that the large urban populations south of the Delta would face acute shortages, despite the fact that senior agricultural water users, especially those in the Sacramento Valley, would have abundant supplies.

In early 1991, this situation prompted the Governor to establish an Emergency Drought Water Bank to arrange short-term transfers of water from willing sellers to meet the "critical water needs" of the purchasers as defined by the rules of bank (Miller, 2000). Recent legislation had set the stage for this program by clarifying rules for temporary transfers of water (Gray, 1990), but privately arranged short-term transfers still could not be approved quickly enough to alleviate the impacts of the drought. A committee of prospective purchasers was established to set the purchase price, which was pegged at $125 per acre-foot (AF) of consumptive use, based on rough estimates that farmers of a variety of field crops could earn slightly more by fallowing their land and selling water at that price than they would earn by using the water to grow a crop. A price escalation clause assured the sellers that they would not lose out if prices rose after they agreed to sell their water. Personnel from the California Department of Water Resources then negotiated the purchase contracts, moved the water through the SWP and coordinated deliveries to the purchasers. The experiment far exceeded expectations, resulting in acquisition of approximately 820 000 AF of water (1 km^3). That substantially exceeded the quantity that prospective buyers had committed to take at the $175 price charged for delivered water, which included pumping costs and payment for a portion of the water to be left in the Sacramento River to help repel the saltwater front. The unsold water was held as carryover storage in SWP facilities, at substantial financial cost to the DWR. As it turned out, 1992 was also very dry, and the state adopted a more cautious approach to buying water, reducing the price offered to sellers to $50/AF and ensuring the commitments of purchasers prior to acquiring the water. The 1991 water bank was California's first foray into large-scale state-managed water marketing, and the lack of previous experience led to unexpected outcomes that provided opportunities to improve practices for subsequent state-run water banks.

The greatest contribution of this experiment was to demonstrate the capacity of water marketing to reduce the adverse impacts of a severe drought by reallocating water to more valuable uses. Experience with the state's drought water banks also dispelled previously widely held fears that water markets would allow the supposedly insatiable water demands of southern California's cities to dry up California's farms. In fact, agricultural districts south of the Delta were major water purchasers, accounting for a quarter of the purchases from the 1991 bank, 71% of the purchases from the 1992 bank, and 85% of the purchases from the 1994 bank (Jercich, 1997). Another major lesson was a better comprehension of the difficulty of verifying the quantity of water actually added to the system as a result of the purchases, given limited documentation of prior water use by the sellers. That, together with conflicting claims about the extent of adverse third-party impacts from the banking operations, led to changes in the rules for the participation of prospective water sellers. In recent years, growing difficulties in reliably moving water south of the Delta during drought periods, more flexible privately arranged transfer agreements, and implementation of other alternatives for building drought resilience have lessened the role for such state-run water banks, and none have been implemented since 2009. Nevertheless, the precedent established by the water banks set the stage for the subsequent significant role of privately arranged water trading as a drought management tool. Water trading expanded significantly in the next two decades, with estimated annual averages of 2.1 million acre-feet (MAF) (2.6 km^3) committed for sale or lease, and 1.4 MAF (1.7 km^3) actually moving between parties during the period 2003–2011 (Hanak and Stryjewski, 2012). These are lower-bound estimates of market activity because they do not include transfers between individuals served by the same water district or wholesale agency, which are not recorded by the SWRCB.

Coincident with the increased role of water marketing, there have been significant changes in the crop mix of Central Valley agriculture. Acreages of water-intensive field crops, including cotton and alfalfa, declined by 550000 acres (222577 hectares) between 2000 and 2011, dropping more precipitously by an additional 930000 acres (376358 hectares) between 2011 and 2014. At the same time, acreages planted to fruit and nut trees have increased by 570000 acres (230671 hectares) since 2000, dominated by increases in high-value almonds and pistachios (Cooley et al., 2015). This shift has been accompanied by a reallocation of water use from lower-valued field crops to these perennial crops, which yield much higher gross revenue per unit of water (Medellin-Azuara and Lund, 2015), but which cannot tolerate significant disruptions in irrigation applications. The capacity of markets to transfer water from fallowed field crops to perennial crops has played an important role, along with increased groundwater use, in mitigating the economic impacts of the recent drought. In fact, inflation-adjusted statewide crop revenues reached record levels in 2013, while agricultural employment peaked in 2014 (Cooley et al., 2015).

The 1987–1992 drought also focused attention on the environmental costs of large-scale pumping operations in the Delta. In addition to actions required by the aforementioned ESA listings, the federal government changed its strategy for management of the CVP with passage of the 1992 Central Valley Project Improvement Act (CVPIA) (Public Law 102-575). These actions have contributed to more frequent reductions in water deliveries to CVP contractors, especially those south of the Delta. The statutory purposes of the CVPIA include restoring and enhancing fish, wildlife and associated habitats. It requires the Bureau of Reclamation to dedicate 800000 AF of water annually to fish and wildlife, provides funding for habitat restoration, and implements changes in contract renewal terms, water transfer rules and pricing practices (Reclamation, 2015b). In 2014 and 2015, no water was delivered to most CVP agricultural contractors, because the limited available supplies were needed to meet environmental obligations and critical urban uses. The SWP also has experienced more frequent difficulties in meeting water delivery contracts, with full allocations delivered to all contractors in only six of the past 25 years (CDWR, 2015).

The declining reliability of water deliveries from these major projects has spurred a number of responses from southern California's water sector. Most importantly, major urban water providers have embarked on a program of demand reduction and diversification of supply sources. The Metropolitan Water District of Southern California is a cooperative water wholesaler, whose 26 member agencies serve approximately 19 million customers. MWD has developed a series of integrated water resources plans since 1996, which call for combining new regional water storage projects, conservation measures, drought-year water transfers, stormwater capture, wastewater recycling, and other activities to improve drought resilience. Since the mid-1990s, MWD has entered into a number of aquifer recharge and recovery agreements, as well as water transfer and exchange contracts, with agricultural water districts in southern California. MWD's ability to draw on those resources significantly reduced the impacts of the most recent drought on urban southern California (MWD, 2015). Examples of innovative water supply agreements include a 35-year contract with the Palo Verde Irrigation District (PVID) under which participating farmers are paid to reduce their water use by not irrigating a portion of their land. The water saved is then made available to MWD. The acreage fallowed increases during drought years, with a maximum of 29% of lands within PVID being fallowed in any given year (MWD, 2010). In 2014, that agreement added 43 010 AF of water (53 Mm^3) to MWD's supplies. From July 2014 to June 2015, other agricultural conservation, canal lining, and aquifer storage and recovery contracts were estimated to have yielded an additional 374 522 AF (462 Mm^3) to MWD and its member agencies. Recycled water production and recovery of previously recharged groundwater in urban aquifers were estimated to have contributed similar volumes to total MWD supplies (MWD, 2015).

MWD's individual member agencies also have implemented their own diversification programs. For example, in December 2015, the Carlsbad Desalination Plant began delivering water to the San Diego County Water Authority. It is the nation's largest and most technologically advanced seawater desalination plant, capable of providing up to 56 000 AF (69 Mm^3) of water per year, or roughly enough to serve 400 000 urban customers (Carlsbad Desalination Plant, 2016). A less expensive alternative has been put in place by the Orange County Water District, which began augmenting its groundwater supplies in 2008 with highly treated wastewater (Orange County Water District, n.d.).

Other innovations in southern California include new approaches to inter-agency coordination on water resource planning and management.

The Santa Ana Water Project Authority (SAWPA) is a notable example. This organization was formed in 1968 to resolve bitter conflicts between several large water districts in the rapidly urbanizing Santa Ana River Basin. In recent years, it has evolved into a model of collaborative integrated water planning that takes a whole-system view of the watershed from its forest-covered headwaters to the Pacific Ocean. This process involves a wide range of water management, environmental and community-based stakeholders and government agencies, and it has led to improvements in water quality, successful water conservation programs and cost-effective infrastructure investments (Cantu, 2016).

The 2012–2015 drought triggered other changes at the state level. In 2014 and 2015, the SWRCB issued curtailment orders for many water rights for the first time since the 1977–1978 drought. As of June 12, 2015 curtailment orders on the Sacramento and San Joaquin Rivers extended to all water rights with priority dates junior to 1903, leaving those rights holders unable to draw upon their usual sources of supply. This blanket approach was necessarily crude, owing to incomplete records on prior uses (Lund et al., 2014). To overcome these data limitations, new legislation (SB 88) was passed in 2015 requiring metering of all surface water diversions exceeding 10 AF (12335 m^3) per year, along with annual reporting of use. In addition, in response to allegations of illegal diversions, the SWRCB issued an order requiring certain riparian and pre-1914 water rights holders in the Sacramento and San Joaquin Basins to report their diversions on a monthly basis (SWRCB Order WR 2015-0002-DWR).

Other important legislation includes the 2014 Sustainable Groundwater Management Act, which replaced the existing system of unregulated groundwater use, in most areas, with a requirement that local groundwater sustainability agencies be formed in all high- and medium-priority basins, as designated by the California Department of Water Resources. These are areas in which groundwater overdraft has led to declining aquifer levels and increased pumping costs or, more seriously, to significant land subsidence. Groundwater extraction has provided an important drought buffer for the agricultural sector (Howitt et al., 2014, 2015), but persistent overdraft would reduce the capacity of that buffer. The Act required formation of local agencies by 2017 in 127 high- and medium-priority groundwater basins, implementation of local basin management plans by 2022, and achievement of long-term groundwater sustainability goals by the 2040s (CDWR, 2014; State of California, 2014). If local agencies fail to perform, the SWRCB is authorized to intervene. The program's decentralized approach recognizes the great heterogeneity of California's groundwater basins, with some facing much more rapid levels of decline than others. In addition, many of the Central Valley's alluvial aquifers can best be managed conjunctively with surface water, being recharged during wet periods and drawn down when droughts reduce surface water availability. Local agencies are thought to be best suited to developing locally appropriate plans for such a management regime. The SWRCB is finalizing the detailed requirements for these planning processes, focusing on monitoring aquifer conditions, water use, installation of new wells and the presence of undesirable conditions, including sustained decline in aquifer levels, land surface subsidence and declining water quality (CDWR, 2016).

Other state actions include the Governor's April 1, 2015 Executive Order issuing mandatory conservation targets for urban water utilities to achieve a 25% statewide reduction in potable urban water use (Brown, 2015). For the most part, these targets were met, with each utility developing its own approach to achieving the targets.

THE PATH AHEAD

California has been ahead of much of the rest of the United States in giving serious consideration to climate change, with managers of water and other natural resources increasingly trying to understand and address its impacts, and the state government establishing ambitious goals for reducing greenhouse gas emissions. Although these efforts already were underway prior to the 2012–2015 drought, the coupling of this drought with unprecedented warmth led many Californians to become more concerned about the possible impacts of global climate change on the state's water resources.

Preparing for the changes in seasonal flow patterns and increasingly extreme droughts and floods will require further reforms. Improved documentation of water rights and monitoring of water use will be needed to avoid wasteful misallocation of this valuable resource and to facilitate the use of water transfers to mitigate the impacts of droughts. Many proposals have been put forward, including streamlining administration of water rights by bringing all surface water users under the SWRCB's permitting system, quantifying groundwater rights in priority basins, modernizing the water accounting system, and improving the transparency and flexibility of water management for environmental protection (Gray et al., 2015).

Also needed will be more effective cooperative approaches to water planning that encompass all of the social and environmental values at stake. California's citizens are making strides in that direction, but conflicting values and cherished preferred solutions still impede the development of solutions to long-recognized water management problems. To make further progress, new approaches that enhance the capacity to overcome such differences while promoting creative problem solving are needed. Participants in future water planning efforts must strive to understand the implications of different plausible trajectories of changing water-related risks and growing pressures on resources. They also should develop a willingness to embrace uncertainty by acknowledging the imperfect predictability of future local-scale climate changes and the responses of complicated water-dependent socioecological systems to those changes. A purposeful strategy of learning from inevitable surprises also will be an important tool for responding to the stresses that a variable and changing climate will produce.

California's experiences during its recent droughts demonstrate the importance of a diversified approach to building water supply security and the importance of harnessing the economic interests of individual water users to promote water use efficiency in the near term and protection of long-term public interests in water resources. California's recent attention to measuring and monitoring surface water use and enlisting local water users in defining groundwater sustainability goals and implementing strategies to achieve them are important steps that could provide models for institutional improvements in other water-limited settings. By pairing such command-and-control measures as mandatory drought-year urban conservation targets with comprehensive attention to balancing protection of ecosystems, creating new water storage, water recycling and stormwater management initiatives, California has adopted an "all-of-the-above" water management strategy (Lohan, 2016). The increased attention given to supporting local water planning processes promises to reduce adverse externalities and ease conflicts while facilitating reallocation of available water supplies to the most socially valuable uses.

NOTES

1. Estimates of the share of agriculture in California's water use differ depending on whether the figure is based on withdrawals or consumptive use (i.e. water withdrawn from a source that does not return to a usable water body after use) and whether instream flows are counted as a water use, or only human water uses are considered. For a summary of different estimates, see Johnson and Cody (2015).
2. The 1922 Colorado River Compact and subsequent legal rulings cap California's annual consumption of water from the Colorado River at 4.4 million acre-feet (5.4 km^3) (Robison and MacDonnell, 2014).

REFERENCES

Adam, J.C., A.F. Hamlet and D.P. Lettenmaier (2009), 'Implications of global climate change for snowmelt hydrology in the twenty-first century', *Hydrological Processes*, **23**, 962–972.

Adams, H.D., C.H. Luce, D.D. Breshears, C.D. Allen, M. Weiler, V.C. Hale, A.M.S. Smith and T.E. Huxman (2012), 'Ecohydrological consequences of drought- and infestation-triggered tree die-off: Insights and hypotheses', *Ecohydrology*, **5**, 145–159.

Allen, C.D., A.K. Macalady, H. Chenchouni, D. Bachelet, N. McDowell, M. Vennetier, T. Kitzberger, A. Rigling et al. (2010), 'A global overview of drought and heat-induced tree mortality reveals emerging climate change risks for forests', *Forest Ecology and Management*, **259** (4), 660–684.

Brown, E.G., Governor of California (2015), Executive Order B-29-15, http://www.waterboards.ca.gov/waterrights/water_issues/programs/drought/docs/040115_executive_order.pdf.

California Climate Tracker (2016), http://www.wrcc.dri.edu/monitor/cal-mon/frames_version.html (accessed 15 January 2016).

Cantu, C. (2016), 'Responses of southern California's urban water sector to changing stresses and increased uncertainty: Innovative approaches', in K.A. Miller, A.F. Hamlet, D.S. Kenney and K.T. Redmond (eds), *Water Policy and Planning in a Variable and Changing Climate*, Boca Raton, FL: CRC Press, pp. 275–288.

Carlsbad Desalination Plant (2016), http://carlsbaddesal.com/ (accessed 18 January 2016).

CDWR (California Department of Water Resources) (2014), *Public Update for Drought Response: Groundwater Basins with Potential Water Shortages, Gaps in Groundwater Monitoring, Monitoring of Land Subsidence and Agricultural Land Fallowing*, November, Sacramento, CA: CDWR, http://www.water.ca.gov/waterconditions/docs/DWR_PublicUpdateforDroughtResponse_GroundwaterBasins.pdf.

CDWR (California Department of Water Resources) (2015), *Drought in California*, Fall, http://www.water.ca.gov/waterconditions/docs/DWR_DroughtBroch_070815-web.pdf.

CDWR (California Department of Water Resources) (2016), *Sustainable Groundwater Management: Draft Groundwater Sustainability Plan Emergency Regulations*, 18 February, http://www.water.ca.gov/groundwater/sgm/pdfs/DRAFT_GSP_Emergency_Regulations_021816.pdf.

Cooley, H., K. Donnelly, R. Phurisamban and M. Subramanian (2015), *Impacts of California's Ongoing Drought: Agriculture*, Oakland, CA: Pacific Institute.

Dettinger, M.D., F.M. Ralph, T. Das, P.J. Neiman and D.R. Cayan (2011), 'Atmospheric rivers, floods, and the water resources of California', *Water*, **3**, 455–478.

Dettinger, M.D., B. Udall and A. Georgakakos (2015), 'Western water and climate change', *Ecological Applications*, **25** (8), 2069–2093.

Gray, B. (1990), 'Water transfers in California: 1981–1989', chap. 2 in R.S. Robinson, L.J. MacDonnell, C.W. Howe, T.A. Rice and M. Squillace, *The Water Transfer Process as a Management Option for Meeting Changing Water Demands*, Vol. II, U.S. Geological Survey Project Report, Grant 14-08-0001-G1538, http://scholar.law.colorado.edu/cgi/viewcontent.cgi?article=1092&context=books_reports_studies.

Gray, B., E. Hanak, R. Frank, R. Howitt, J. Lund, L. Szeptycki and B. Thompson (2015), *Allocating California's Water: Directions for Reform*, San Francisco, CA: Public Policy Institute of California.

Hanak, E. and E. Stryjewski (2012), *California's Water Market, by the Numbers: Update 2012*, San Francisco: Public Policy Institute of California.

Hanemann, M., C. Dykman and D. Park (2015), 'California's flawed surface water rights', in Allison Lassiter (ed.), *Sustainable Water Challenges and Solutions from California*, Berkeley: University of California Press, pp. 52–81.

Healey, M., M. Dettinger and R. Norgaard (eds) (2008), *The State of Bay–Delta Science, 2008*, CALFED Science Program, http://science.calwater.ca.gov/publications/sbds.html.

Howitt, R., J. Medellín-Azuara, D. MacEwan, J. Lund and D. Sumner (2014), 'Economic analysis of the 2014 drought for California agriculture', Center for Watershed Sciences, University of California, Davis, UC Agricultural Issues Center, and ERA Economics, Davis, CA, https://watershed.ucdavis.edu/files/biblio/DroughtReport_23July2014_0.pdf.

Howitt, R.E., D. MacEwan, J. Medellín-Azuara, J.R. Lund and D.A. Sumner (2015), 'Economic analysis of the 2015 drought for California agriculture', Center for Watershed Sciences, University of California, Davis, https://watershed.ucdavis.edu/files/biblio/Economic_Analysis_2015_California_Drought__Main_Report.pdf.

Jercich, S. (1997), 'California's 1995 Water Bank Program: Purchasing water supply options', *Journal of Water Resources Planning and Management*, **123** (1), 59–65.

Johnson, Renée and Betsy A. Cody (2015), *California Agricultural Production and Irrigated Water Use*, CRS Report No. R44093, Washington, DC: Congressional Research Service, June, https://www.fas.org/sgp/crs/misc/R44093.pdf.

Jones, J. (2015), 'California's most significant droughts: Comparing historical and recent conditions', Report, State of California, Department of Water Resources, February, http://www.water.ca.gov/waterconditions/docs/California_Signficant_Droughts_2015_small.pdf.

Jones, J. (2016), 'California, a state of extremes: Management framework for present-day and future hydroclimate extremes', in K.A. Miller, A.F. Hamlet, D.S. Kenney and K.T. Redmond (eds), *Water Policy and Planning in a Variable and Changing Climate*, Boca Raton, FL: CRC Press, pp. 207–222.

Lindström, S. (1990), 'Submerged tree stumps as indicators of mid-Holocene aridity in the Lake Tahoe region', *Journal of California and Great Basin Anthropology*, **12** (2), 146–157.

Lohan, T. (2016), 'Felicia Marcus: California's drought lessons', *Water Deeply* online report, Interview with Felicia Marcus, Chair, State Water Resources Control Board, 17 March, http://www.waterdeeply.org/articles/2016/03/9875/felicia-marcus-californias-drought-lessons/.

Luce, C., P. Morgan, K. Dwire, D. Isaak, Z. Holden and B. Rieman (2012), *Climate Change, Forests, Fire, Water, and Fish: Building Resilient Landscapes, Streams, and Managers*, General Technical Report RMRS-GTR-290, Fort Collins, CO: U.S. Department of Agriculture, Forest Service, Rocky Mountain Research Station.

Lund, J., B. Lord, W. Fleenor and A. Willis (2014), 'Drought curtailment of water rights – problems and technical solutions', Center for Watershed Sciences Report, University of California, Davis.

Medellín-Azuara, J. and J. Lund (2015), 'Dollars and drops per California crop', *CaliforniaWaterBlog.com*, posted 14 April by UC Davis Center for Watershed

Sciences, https://californiawaterblog.com/2015/04/14/dollars-and-drops-per-crop-in-california/.
Meko, D.M., C.A. Woodhouse and R. Touchan (2014), 'Klamath/San Joaquin/Sacramento hydroclimatic reconstructions from tree rings', Report to the California Department of Water Resources, Agreement 4600008850.
Miller, K.A. (2000), 'Managing supply variability: The use of water banks in the western United States', in D.A. Wilhite (ed.), *Drought: A Global Assessment*, Vol. II, London: Routledge, pp. 70–86.
Mount, J., E. Freeman and J. Lund (2014), *Water Use in California*, San Francisco, CA: Public Policy Institute of California (PPIC), http://www.ppic.org/main/publication_show.asp?i=1108.
MWD (Metropolitan Water District of Southern California) (n.d.), 'Imported supplies', http://www.mwdh2o.com/AboutYourWater/Sources%20Of%20Supply/Pages/Imported.aspx (accessed 15 January 2016).
MWD (Metropolitan Water District of Southern California) (2010), *Integrated Water Resources Plan*, Report No. 1373, October, Los Angeles, CA: MWD, http://www.mwdh2o.com/Reports/2.4.1_Integrated_Resources_Plan.pdf.
MWD (Metropolitan Water District of Southern California) (2015), *Annual Report for the Fiscal Year July 1, 2014 to June 30, 2015*, Los Angeles, CA: MWD, http://www.mwdh2o.com/PDF_Who_We_Are/2015-MWD-Annual-Report-web.pdf.
NOAA (National Oceanic and Atmospheric Administration) National Centers for Environmental Information (2015), *Climatological Rankings*, California, http://www.ncdc.noaa.gov/temp-and-precip/climatological-rankings/?parameter=pcp&state=004&div=00&month=oct&year=2015 (accessed 28 November 2015).
Norgaard, R.B. (2016), 'California's Sacramento–San Joaquin Delta: Reflections on science, policy, institutions, and management in the Anthropocene', in K.A. Miller, A.F. Hamlet, D.S. Kenney and K.T. Redmond (eds), *Water Policy and Planning in a Variable and Changing Climate*, Boca Raton, FL: CRC Press, pp. 223–238.
Orange County Water District (n.d.), Groundwater Replenishment System, http://www.ocwd.com/what-we-do/water-reuse/.
Reclamation (2015a), 'About the Central Valley Project', http://www.usbr.gov/mp/cvp/about-cvp.html (accessed 1 December 2015).
Reclamation (2015b), 'Central Valley Project Improvement Act (CVPIA)', http://www.usbr.gov/mp/cvpia/ (accessed 1 December 2015).
Robison, J.A. and L.J. MacDonnell (2014), '*Arizona v. California* & the Colorado River Compact: Fifty years ago, fifty years ahead', *Arizona Journal of Environmental Law and Policy*, **4**, 130–159.
Roos, M. and S. Sahota (2012), 'Contrasting snowpack trends in the Sierra Nevada of California', Paper presented at the 80th Western Snow Conference, Anchorage, AK, May, http://www.westernsnowconference.org/sites/westernsnowconference.org/PDFs/2012Roos.pdf (accessed 3 December 2015).
State of California (2014), Sustainable Groundwater Management Act, AB 1739, SB 1319 and SB 1168, http://groundwater.ca.gov/legislation.cfm.
State of California (2015), *Management of the State Water Project*, Bulletin 132-13, April, Sacramento: California Department of Water Resources, http://www.water.ca.gov/swpao/docs/bulletins/bulletin132/Bulletin132-13.pdf.
Stewart, I.T. (2009), 'Changes in snowpack and snowmelt runoff for key mountain regions', *Hydrological Processes*, **23**, 78–94.
USDA (U.S. Department of Agriculture) (2012), 'Ranking of market value of

ag products sold, California', *2012 Census of Agriculture*, USDA National Agricultural Statistics Service, http://www.agcensus.usda.gov/Publications/2012/Online_Resources/Rankings_of_Market_Value/California/index.asp (accessed 25 November 2015).

USDA-ERS (U.S. Department of Agriculture – Economic Research Service) (2015), *Agricultural Production and Prices*, http://www.ers.usda.gov/data-products/ag-and-food-statistics-charting-the-essentials/agricultural-production-and-prices.aspx (accessed 25 November 2015).

U.S. Drought Monitor (2015), 'California', National Drought Mitigation Center, http://droughtmonitor.unl.edu/Home/StateDroughtMonitor.aspx?CA (accessed 25 November 2015).

USGS (United States Geological Survey) (2015), 'Annual runoff estimate for California', Time series 1901–2015, http://ca.water.usgs.gov/data/drought/runoff.html (accessed 25 November 2015).

Water Education Foundation (n.d.), 'Sacramento–San Joaquin Delta chronology', http://www.watereducation.org/aquapedia/sacramento-san-joaquin-delta-chronology (accessed 30 December 2015).

Western Regional Climate Center (2015), Map of California average annual precipitation, http://www.wrcc.dri.edu/pcpn/ca.gif (accessed 1 December 2015).

Zellmer, S.B. and C.A. Klein (2016), 'Floods as unnatural disasters: The role of law', in K.A. Miller, A.F. Hamlet, D.S. Kenney and K.T. Redmond (eds), *Water Policy and Planning in a Variable and Changing Climate*, Boca Raton, FL: CRC Press, pp. 361–374.

11. Building urban climate resilience in Vietnam and Bangladesh

Craig Johnson, Iftekharul Haque, Yvonne Su and Kristy May

INTRODUCTION

There is now a growing recognition that cities are becoming increasingly vulnerable to the effects of climate change (McGranahan et al., 2007; Revi et al., 2014; Johnson et al., 2015; WWF, 2015). According to the United Nations, the world's urban population is expected to grow by another 3 billion people by the year 2050, putting unprecedented pressure on the world's terrestrial, hydrological and atmospheric systems, as well as essential urban services (UN/DESA, 2012: 18). Of the 63 urban areas with populations greater than 5 million, 39 are located in areas that have a high risk of flooding, cyclones or drought, 72 per cent are located 'on or near the coast', and two-thirds are in Asia. Although the highest rates of urbanization have been happening in Latin America, the vast majority of future urban growth is expected to occur in Asia and sub-Saharan Africa (UN/DESA, 2012).

In light of these trends, increasing attention is now being paid to the challenge of building the resilience of urban populations, systems, assets and infrastructure to extreme climate events (Romero-Lankao and Dodman, 2011; Atkins, 2012; Revi et al., 2014; Rosenzweig and Solecki, 2014). The IPCC's Fifth Assessment Report (Revi et al., 2014: 541), for instance, highlights 'the importance of city and municipal governments acting now to incorporate climate change adaptation into their development plans and policies and infrastructure investments'. Building urban resilience to climate change is therefore a pressing global priority.

This chapter explores the challenges of building climate resilience in two of the world's most vulnerable city regions: Dhaka, Bangladesh and Ho Chi Minh City (HCMC), Vietnam (Table 11.1). According to one recent global assessment, HCMC was ranked fourth in the world for annual and projected flood losses (relative to GDP); the Dhaka region was number eight (Hallegatte et al., 2013). According to another study (McGranahan et al., 2007: 26), 55 per cent of Vietnam's population and 46 per cent of Bangladesh's population are in areas classified as low-elevation coastal zones (defined as the contiguous areas along the coast that are less than 10 metres above sea level).

According to the United Nations Office for Disaster Risk Reduction (UNISDR, 2012: 11), 'a resilient city is characterized by its capacity to withstand or absorb the impact of a hazard through resistance or adaptation, which enable it to maintain certain basic functions and structures during a crisis, and bounce back or recover from an event'. However, building resilience entails an ability to overcome the rigidities put in place by years of investment (or the lack thereof) in particular modes of sanitation, transportation, housing and flood control (see World Bank, 2010; Revi et al., 2014). Owing to their low elevation, poor drainage and exposure to storm surges, coastal cities are often highly vulnerable to urban flooding, waterlogging and the spread of waterborne disease (Revi et al., 2014).

Recent empirical work on the factors affecting urban climate adaptation has shown that cities can build resilience to short-term disasters and longer-term climatic changes when they are 'nested' within national systems that specify the rights of local municipal bodies to enact plans and policies governing climate sensitive sectors like waste and water treatment, sanitation, land use planning, and emergency services (Tanner et al., 2009; Gore, 2015; Johnson et al., 2015). However, cities are often poorly placed when it comes to negotiating powers of legislation, regulation and taxation, highlighting the multi-scalar disparities that frequently divide cities and nation-states (Gore, 2015; Johnson et al., 2015). Moreover, cities are often under great pressure to facilitate the conversion of wetlands, water bodies and other ecologically sensitive areas for the purposes of commercial, residential and industrial development (Satterthwaite et al., 2010; Boyd et al., 2015; Gore, 2015; Johnson et al., 2015).

In what follows, we argue that national adaptation planning in both countries has been framed primarily in relation to the needs of rural areas, relegating cities to a relatively marginal status. Far from mainstreaming climate change resilience, urban planning in Vietnam and Bangladesh has instead facilitated the conversion of ecologically sensitive wetlands and water bodies, often at the expense of the urban poor. In both cases, national and sub-national authorities have used urban and peri-urban land as a means of facilitating investment, thereby restructuring the role of domestic and international capital in a rapidly globalizing urban economy.

Our empirical analysis is based on six months of primary field research in Dhaka and HCMC, which involved interviews with academics, urban planners, non-governmental organizations and private sector land developers,

Table 11.1 Dhaka and Ho Chi Minh City at a glance

	Dhaka, Bangladesh	Ho Chi Minh City, Vietnam
Status	Capital of Bangladesh; headquarters of Dhaka Division and District of Dhaka	City-province (HCMC and four other cities in Vietnam have provincial status)
Area (square kilometres)	300.97	2096
Population (in millions)	9.25 (2011) (most populous)	8.22 (2015) (most populous)
Population density (per square kilometre)	30 748.8 (2011)	3900 (2015)
Elevation above sea level	13 ft (4 m)	62 ft (19 m)
Poverty rate	15.7% (2010)	0.31% (2010)

Note: Years in parentheses denote the survey or census year.

Source: BBS (2015); GSOV (2016).

as well as a systematic review of the major legislative and policy documents relating to urban planning and climate change adaptation in both cities.

DHAKA, BANGLADESH

Dhaka's Climate Change Vulnerabilities

According to WWF (2015), Dhaka is one of the most vulnerable cities in Asia to the impacts of climate change. The study estimates that Dhaka has a relatively higher exposure to environmental disasters (8 out of 10) and higher socio-economic sensitivity (8 out of 10) with lowest adaptive capacity (10 out of 10 in inverse adaptive capacity) among the major cities of Asia. The influx of migrants and unplanned growth of the city made the city more vulnerable to climate change related shocks.

A number of factors are at play here. First, urban encroachment into wetlands and low-elevation areas has reduced the drainage capacity of the built environment, further exposing urban populations to waterlogging and extreme flood events (McGranahan et al., 2007). Second, overextraction of groundwater and reduced runoff from the Himalayas have depleted Dhaka's groundwater level, reducing access to clean water. Third, Dhaka is highly exposed to windstorms and flooding as a result of the annual monsoons, whose impacts are expected to worsen with climate change (Revi et al., 2014). Finally, the city is growing rapidly at a rate of 4 per cent per year in the 2000s, and the city's population is expected to grow to 20 million by the year 2025.

Without a doubt, those most vulnerable to climate change are the estimated 5 million people living in the informal settlements classified as urban slums, where the inhabitants often live and work in close proximity to wetlands, water bodies and canals, and which are notoriously ill provisioned in terms of access to clean water and sanitation services. The Bangladesh Centre for Urban Studies' census (CUS, 2006) found that around 60 per cent of slums in Dhaka have little or no drainage and are vulnerable to frequent flooding. In addition, the housing materials of more than one-third of Dhaka's population were too vulnerable to withstand the impact of large-scale environmental disasters.

Adaptation at the National Level: Bangladesh's Legislative Framework on Climate Change

In the absence of a formal urban adaptation plan, there are a number of national adaptation policies that have relevance to cities in Bangladesh. The Bangladesh Climate Change Strategy and Action Plan (BCCSAP) was first released in 2009 (GoB, 2009), and has since been adopted as the principal guiding document for climate change adaptation planning in Bangladesh (Ayers et al., 2014). At the heart of the plan are six pillars: food security, social protection and health; comprehensive disaster management; infrastructure; research and knowledge management; mitigation and low-carbon development; and capacity building and institutional strengthening. However, the BCCSAP has been criticized for focusing too heavily on rural poverty and vulnerability, underplaying the impact of climate change on cities and low-income urban populations (Banks et al., 2011). Of the 44 programmes being supported by the BCCSAP, only two directly address climate change vulnerability in the urban context.

A second planning document is the Sixth Five Year Plan (which ran from 2011 to 2015) (GoB, 2011). At the heart of this plan is an effort to mainstream climate change into national planning, by integrating environmental, climate change and disaster risk reduction considerations into project design, budgetary allocations and implementation of plans and policies. Based on the framework provided by the GoB (2009), the plan supports 35 programmes aimed at addressing vulnerabilities related to environmental hazards and climate change. However, much like the BCCSAP, only two of these programmes are explicitly aimed at addressing climate change vulnerabilities in cities.

Underlying the propensity of national adaptation plans to emphasize rural adaptation at the expense of urban priorities and concerns is a tendency on the part of foreign aid donors to prioritize rural development in their own adaptation programming. Of the USD 72 million that supported 15 donor-funded national adaptation programmes of action (NAPA)[1] projects between 2005 and 2008, only 2.75 per cent was allocated for cities. This tendency of rural bias in climate change project funding is also found in the projects of the Bangladesh Climate Change Resilience Fund. As of 2015, the Fund had allocated USD 146 million for 13 projects. However, only one project (USD 0.5 million) directly addresses vulnerability in urban areas (BCCRF, 2015).

Adaptation at the Urban Scale: Urban Planning and Governance Capacity

At the sub-national level, there is also a disconnect between adaptation efforts and urban planning. Neither of Dhaka's two city corporations[2] has policy documents aimed explicitly at addressing or reducing urban vulnerability to climate change. In 2011, Dhaka North City Corporation established a Climate Change, Environment and Disaster Management Department, which is responsible for administering small-scale adaptation projects. However, neither of the city corporations has the power to regulate land and industrial development in Dhaka (Araos, 2015). On the contrary, an autonomous government agency by the name of RAJUK[3] holds primary responsibility for regulating, approving and promoting commercial and industrial development in Dhaka.

Arguably the most important aspect of RAJUK's authority rests in its ability to prepare, implement and enforce the comprehensive land use plans that have been developed for Dhaka and surrounding areas. In 1995, RAJUK took the lead in preparing the Dhaka Metropolitan Development Plan (DMDP), a comprehensive land use policy that later formed the basis for the Detailed Area Plan (DAP), which was approved in 2010. In theory, the DMDP contained a number of provisions aimed at addressing the factors contributing to urban climate vulnerability, including protection for green belts, water bodies, flood zones, wetlands and agricultural land in and around the city. In practice, the ability of the planning document to regulate land use was undermined by a weak governance regime and a powerful constellation of politicians and private sector land developers. Perhaps the most illustrative example of the government's inability to regulate its own land use policies came in the late 1990s, when RAJUK approved plans to develop a 6000-acre township, known as Purbachal New Town, in the middle of an area that it had itself designated as a flood flow zone in the DMDP.

The inability of the government to regulate and enforce its own land use policies reflects the demand for urban land in Bangladesh and the limited autonomy of government in relation to private sector land developers. Driving the rapid economic and geographical expansion of Dhaka has been the growth of the ready-made garment (RMG) sector, an industry that employs an estimated 4 million people. According to one study (Khandker and Mahmud, 2012), 80 per cent of the country's 5000 RMG factories are located in Dhaka, attracting large numbers of temporary and permanent migrants from all over Bangladesh. According to another, by the Real Estate Housing Association of Bangladesh (REHAB), housing rent in Dhaka increased at an annual rate of 250 per cent between 1991 and 2007 (REHAB, 2012). Thus it is unsurprising that regulations aimed at protecting wetlands, flood zones, water bodies and small-scale agricultural activities have been ignored in favour of high-end commercial and residential real estate development (Habib, 2015).

In short, Dhaka lacks a comprehensive climate change adaptation strategy that may be used as the basis for building urban resilience to climate change. The BCCSAP and Sixth Five Year Plan have provisions for climate change adaptation, but neither puts any significant emphasis on urban areas. Notwithstanding the linkages that can be made between rural vulnerability and urban migration, cities do not feature prominently in either of these planning documents. At the municipal level, the DMDP and the DAP contain provisions for protecting flood zones, wetlands and water bodies, but these are summarily ignored, further exacerbating the vulnerability of urban populations to waterlogging, windstorms and extreme flood events.

HO CHI MINH CITY, VIETNAM

Ho Chi Minh City's Climate Change Vulnerabilities

Located in the delta created by the Dong Nai and Saigon rivers, HCMC is highly vulnerable to the effects of climate change. Most concerning are the effects of rising sea levels, saltwater intrusion, changing precipitation patterns, more frequent tropical storms, and urban heat island effects (Webster and McElwee, 2009: 3). In 2011, 1.9 million people in HCMC were directly exposed to coastal flooding. By 2070, this number is expected to rise to 9.2 million (Hanson et al., 2011: 100). Further, a 1-metre rise in sea levels is predicted to inundate 43 per cent of HCMC's total land area (Carew-Reid, 2008: 2). As HCMC is an emerging megacity, rapid urbanization and the influx of migrants further exacerbate its vulnerability to climate change.

From 2000 to 2012, agricultural land in HCMC fell from 130729 to 117625 ha (GSO, 2013), with much of this loss occurring in low-elevated areas on the urban periphery. Owing to rigid urban plans and inaccurate government data used for planning projections, much of the urban development occurs in unplanned and unauthorized urban areas (Coulthart et al., 2006). As a result, many residential areas in HCMC have been built in an ad hoc manner, often lacking proper drainage and sewer systems (Coulthart et al., 2006: 32).

Similar to Dhaka, HCMC is the largest and fastest growing city in the country, with its population increasing by 3.9 million between 1989 and 2012, and population densities reaching as high as 45 428 people/km^2 in some districts (GSO, 2013). However, as a result of Vietnam's long-maintained household registration system, many households occupy land in HCMC illegally. Official data on HCMC's population often exclude the high levels of temporary workers and unregistered migrants in the city, with the 'floating population' estimated to be an additional 2 million people (Jones, 2008: 44).

Unregistered migrants are often highly vulnerable to the effects of rapid-onset disasters and long-term climate change. Lacking formal registration, migrant populations typically inhabit densely populated districts located in low-lying areas (Webster and McElwee, 2009: 4). Similar to Dhaka, these are the areas that are most vulnerable to flooding, because they are built on wetlands or located in areas prone to tidal flooding lacking adequate drainage. Further, the precarious legal status of the unregistered migrants leaves them unable to legally purchase land or build a house, register children in school, or access public services (Pincus and Sender, 2008: 117). Over 90 per cent of migrants live in rented housing, temporary shelters or other forms of shelter (Haugton et al., 2010: 127).

Adaptation at the National Level: Vietnam's Legislative Framework on Climate Change

Following its ratification of the United Nations Framework Convention on Climate Change (UNFCCC) in 1994 and the Kyoto Protocol in 2002, the Vietnamese government has introduced a number of national strategies aimed at addressing climate change, including Vietnam's Sustainable Development Strategy (2004), the National Strategy for Natural Disaster Prevention, Response and Mitigation (2007), the National Target Programme to Respond to Climate Change (2008), the National Strategy on Climate Change (2011) and the accompanying National Action Plan to Respond to Climate Change for 2012–2020 (2012). Similar to Bangladesh, none of the above climate change strategies has a specific urban focus. Further, many of the priorities related to urban areas focus only on reducing physical vulnerabilities to climate change. The National Strategy on

Climate Change, for instance, highlights the need to adjust socio-economic development strategies and master plans to consider risks and uncertainties of climate change and sea-level rise, and integrate climate change into local and regional master planning. The accompanying Action Plan emphasizes the need to reduce flooding in large urban areas, whether through improving dykes and reservoirs or developing irrigation plans. Absent from these climate change plans, however, is a discussion of vulnerable urban populations.

National policy documents also fail to address issues surrounding migration, displacement and urbanization in the context of climate change. From our reading of the National Strategy on Climate Change, the strategy makes no reference to 'migration' or 'displacement'. While the strategy contains a focus on planning residential areas and developing infrastructure in response to climate change, it is uncertain how migrants and future climate migrants will be considered or incorporated into the master planning scheme. Similarly, the National Strategy for Natural Disaster Prevention, Response and Mitigation identifies the relocation of people in disaster prone areas as a central objective, but it fails to consider the vulnerability of unregistered migrants resulting from a rigid household registration system.

Adaptation at the Urban Scale: Urban Planning and Governance Capacity

At the national level, Vietnam has multiple strategies aimed at tackling the effects of climate change. However, comparatively less has been done at the urban scale. The Vietnam Climate Adaptation Partnership (VCAPS),[4] a joint bilateral initiative between Vietnam and the Netherlands, identifies a number of shortcomings in the existing HCMC Spatial Master Plan: 'the master plan does not specifically deal with flood protection. It is unclear which zones are protected to what level and what is expected of developers when building in specific zones' (VCAPS, 2013: 49). Similarly, the German-funded Future Cities programme advises against HCMC's existing plans to rezone agricultural land located in floodplains for residential or industrial development (Storch and Downes, 2012). Using urban development and sea-level rise scenarios, research from this project concluded that the largest driver of future flood risk in HCMC is the urban expansion into 'known flood prone areas' (Storch and Downes, 2011: 524).

Driving the conversion of ecologically sensitive water bodies and wetlands has been the process of market liberalization that has been unfolding since the early 1990s. This is most clearly illustrated in two of HCMC's flagship urban land development projects, Thu Thiem New Urban Area (657 ha) and Saigon South New Urban Area (2975 ha).

Similar to Purbachal New Town in Dhaka, both developments were led by autonomous state agencies that played a key role in attracting foreign and domestic investment in high-end real estate.

In light of the aforementioned vulnerabilities, the state's involvement in such activities appears at odds with its tasks to effectively plan for climate change. Previous recommendations from the HCMC Urban Planning Institute (UPI), a city planning agency, were to develop in a north-west direction on higher-elevated land. However, successful greenfield land development projects, such as those in Saigon South, and soaring land prices in the south end of the city have led to adjustments in the HCMC master plan. In spite of the UPI's recommendations, the government has instead developed plans to expand the city in the southern direction towards the coast.

To summarize, Vietnam has a national policy framework that highlights the importance of mainstreaming climate change in local and regional planning, but HCMC suffers many of the same governance challenges facing Dhaka. Included here are weak and at times contradictory planning documents, enormous pressure to develop wetlands, water bodies and greenfields, and a state that has at times struggled to establish and maintain autonomy from private sector land developers and foreign capital. While initiatives such as VCAPS and the Future Cities programme have been successful in assisting HCMC to identify geographically vulnerable areas of the city and suggestions for adaptation, it is unlikely that these projects will have a discernible effect on the forces that are now driving urbanization and the conversion of ecologically sensitive areas in and around HCMC.

CONCLUSIONS

This chapter has explored the challenges of building climate change resilience in two of the world's most vulnerable city regions. As noted in the introduction to the chapter, the IPCC's Fifth Assessment Report highlights the importance of national and municipal governments incorporating climate change adaptation into their 'development plans and policies and infrastructure investments' (Revi et al., 2014: 541). However, the two cases considered in this chapter highlight the challenges of crafting multi-level policies and solutions that can be used to address the combined impacts of rapid and unplanned urbanization and climate change. Principal among these are the challenges of understanding and ideally addressing the socio-economic determinants of climatic vulnerability at the urban scale.

By way of conclusion, this section now highlights the factors affecting climatic vulnerability, outlining a number of insights about the challenge of building urban resilience to climate change. One of the principal themes that emerges from the comparison is the multidimensional nature of socio-economic and ecological vulnerability in low-elevation coastal cities. Underlying the vulnerability being documented in Dhaka and HCMC are complex processes of rural-to-urban migration, unregulated land use practices and inadequate access to basic urban services. By far the most vulnerable populations in the two cases are the temporary and unregistered migrants, whose lack of access to clean water, sanitation and other essential urban services makes them particularly vulnerable to climate change. Here the task of documenting and supporting the rights of itinerant populations presents its own set of challenges, as does the task of building resilience through the provision of urban services in health, sanitation and disaster risk reduction, aspects of urban governance that have been notably absent in the national adaptation planning strategies of Vietnam and Bangladesh.

A related theme concerns the governance capacity of urban planning institutions. Recent empirical work on the factors affecting effective urban climate governance has shown that cities can incorporate climate change vulnerability into their operational planning and practices, but this depends on cooperation and (ideally) support from national government and international institutions (Johnson et al., 2015). Here the comparison of Dhaka and HCMC suggests that the national government played a more prominent role in Vietnam than was the case in Bangladesh, highlighting (potentially) the historical legacy of central state planning in Vietnam. At the same time, national and municipal authorities in both countries were clearly constrained by a lack of capacity and authority to act beyond the urban scale, reflecting the multi-scalar dimensions of globalization, rural-to-urban migration and climate change.

A third and related theme concerns the ability of cities to govern and adapt to a rapidly changing global and regional economic environment. Here the impacts of liberalization and global economic restructuring loom large, suggesting that the incorporation of Dhaka and HCMC into global economic production systems has created new vulnerabilities that existing authorities and institutions are unable to govern effectively. In both cases, national and municipal authorities have used land as a means of attracting and restructuring foreign and domestic capital, but here it is clear that such processes have come at the expense of ecologically sensitive wetlands and water bodies, displacing tens of thousands of resource dependent communities. Here it is worth emphasizing too that liberalization in both of these contexts has been dependent upon the use of temporary labour working in unsafe conditions, further exacerbating the vulnerability of poor and itinerant populations to climate related shocks and stressors. Whether national and municipal authorities can build the resilience of urban populations, assets and infrastructure to future climatic changes is therefore an open question, reflecting the complexity of urbanization, the capacity of government (at multiple levels) and the impact of global and regional production systems on urban social and ecological systems.

NOTES

1. NAPA is precursor of the BCCSAP. Before BCCSAP, climate change related programmes were guided by NAPA.
2. Dhaka has two city corporations: Dhaka North City Corporation and Dhaka South City Corporation.
3. RAJUK is the abbreviation for Rajdhani Unnayan Corporation in Bengali, which could be translated as Capital Development Authority.
4. VCAPS is a consortium of Dutch experts who provide technical assistance to the HCMC People's Committee with the objective 'to enable and guide the long-term sustainable socio-economic development of HCMC towards the sea, taking into account the effects of climate change' (VCAPS, 2013: 10).

REFERENCES

Araos, M. (2015), 'Adaptation to climate change: How well is Dhaka doing?', *Dhaka Tribune*, 19 September.

Atkins (2012), *Future Proofing Cities: Risks and Opportunities for Inclusive Urban Growth in Developing Countries*, Report, London: Atkins PLC in collaboration with UKAid/DFID and the Development Planning Unit (DPU), University College London.

Ayers, J., S. Huq, H. Wright, A. Faisal and S. Hussain (2014), 'Mainstreaming climate change adaptation into development in Bangladesh', *Climate and Development*, **6** (4), 293–305.

Banks, N., M. Roy and D. Hulme (2011), 'Neglecting the urban poor in Bangladesh: Research, policy and action in the context of climate change', *Environment and Urbanization*, **23** (2), 487–502.

BBS (Bangladesh Bureau of Statistics) (2015), *2014 Statistical Yearbook of Bangladesh*, Dhaka: BBS.

BCCRF (Bangladesh Climate Change Resilience Fund) (2015), *Projects: Bangladesh Climate Change Resilience Fund*, http://bccrf-bd.org/Project.html.

Boyd, E., A. Ghosh and M. Boykoff (2015), 'Climate change adaptation in Mumbai, India', in Craig Johnson, Noah Toly and Heike Schroeder (eds), *The Urban Climate Challenge: Rethinking the Role of Cities in the Global Climate Regime*, New York: Routledge, pp. 139–155.

Carew-Reid, J. (2008), *Rapid Assessment of the Extent and Impact of Sea Level Rise in Viet Nam*, Brisbane, Australia: International Centre for Environmental Management (ICEM).

Coulthart, A., N. Quang and H. Sharpe (2006), *Urban Development Strategy: Meeting the Challenges of Rapid Urbanization and the Transition to a Market Oriented Economy*, Hanoi: World Bank in Vietnam.

CUS (Centre for Urban Studies) (2006), *Slums of Urban Bangladesh: Mapping and Census, 2005*, Dhaka: United States Agency for International Development.
GoB (2009), *Bangladesh Climate Change Strategy and Action Plan 2009*, Dhaka: Ministry of Environment and Forests, Government of the People's Republic of Bangladesh.
GoB (2011), *Sixth Five Year Plan: FY2011–FY2015*, Dhaka: Planning Commission, Government of the People's Republic of Bangladesh.
Gore, C. (2015), 'Climate change adaptation and African cities: Understanding the impact of government and governance on future action', in C. Johnson, N. Toly and H. Schroeder (eds), *The Urban Climate Challenge: Rethinking the Role of Cities in the Global Climate Regime*, New York: Routledge, pp. 205–226.
GSO (2013), *Statistical Yearbook of Ho Chi Minh City 2012*, Ho Chi Minh City: Ho Chi Minh City Statistical Office, Statistical Publishing House.
GSOV (General Statistics Office of Vietnam) (2016), https://www.gso.gov.vn/Default_en.aspx?tabid=491 (accessed 16 February 2016).
Habib, I. (2015), 'Dhaka Structure Plan 2016–35: Prioritise people, not coteries', *Daily Star*, 16 September.
Hallegatte, S., C. Green, R.J. Nicholls and J. Corfee-Morlot (2013), 'Future flood losses in major coastal cities', *Nature Climate Change*, **3** (9), 802–806.
Hanson, S., R. Nicholls, N. Ranger, S. Hallegatte, J. Corfee-Morlot and C. Herweijer (2011), 'A global ranking of port cities with high exposure to climate extremes', *Climatic Change*, **104**, 89–111.
Haugton, J., L. Loan and N. Linh (2010), *Urban Poverty Assessment in Ha Noi and Ho Chi Minh City*, Hanoi: UNDP Viet Nam.
Johnson, C., N. Toly and H. Schroeder (eds) (2015), *The Urban Climate Challenge: Rethinking the Role of Cities in the Global Climate Regime*, New York: Routledge.
Jones, G. (2008), 'Comparative dynamics of six mega-urban regions', in G. Jones and M. Douglass (eds), *Mega-Urban Regions in Pacific Asia: Urban Dynamics in a Global Era*, Singapore: NUS Press, pp. 41–60.
Khandker, S. and W. Mahmud (2012), *Seasonal Hunger and Public Policies: Evidence from Northwest Bangladesh*, Washington, DC: World Bank.
McGranahan, G., D. Balk and B. Anderson (2007), 'The rising tide: Assessing the risks of climate change and human settlements in low elevation coastal zones', *Environment and Urbanization*, **19** (1), 17–37.
Pincus, J. and J. Sender (2008), 'Quantifying poverty in Viet Nam: Who counts?', *Journal of Vietnamese Studies*, **3** (1), 108–150.
REHAB (2012), *A Comprehensive Study on the Real Estate Sector of Bangladesh*, Dhaka: REHAB.
Revi, A., D. Satterthwaite, F. Aragón-Durand, J. Corfee-Morlot, R. Kiunsi, M. Pelling, D. Roberts and W. Solecki (2014), 'Urban areas', in C. Field, V. Barros, D. Dokken, K. Mach, M. Mastrandrea, T. Bilir, M. Chatterjee, K. Ebi et al. (eds), *Climate Change 2014: Impacts, Adaptation, and Vulnerability, Part A: Global and Sectoral Aspects: Contribution of Working Group II to the Fifth Assessment Report of the Intergovernmental Panel on Climate Change*, Cambridge, UK: Cambridge University Press, pp. 535–612.
Romero-Lankao, P. and D. Dodman (2011), 'Cities in transition: Transforming urban centers from hotbeds of GHG emissions and vulnerability to seedbeds of sustainability and resilience', *Current Opinion in Environmental Sustainability*, **3**, 113–120.
Rosenzweig, C. and W. Solecki (2014), 'Hurricane Sandy and adaptation pathways

in New York: Lessons from a first-responder city', *Global Environmental Change*, **28**, 395–408.

Satterthwaite, D., G. McGranahan and C. Tacoli (2010), 'Urbanization and its implications for food and farming', *Philosophical Transactions of the Royal Society B*, **365**, 2809–2820.

Storch, H. and N. Downes (2011), 'A scenario-based approach to assess Ho Chi Minh City's urban development strategies against the impact of climate change', *Cities*, **28**, 517–526.

Storch, H. and N. Downes (eds) (2012), *Land-Use Planning Recommendations: Adaptation Strategies for a Changing Climate in Ho Chi Minh City: Summary for Decision-Makers*, Berlin: German Federal Ministry of Education and Research.

Tanner, T., T. Mitchell, E. Polack and B. Guenther (2009), *Urban Governance for Adaptation: Assessing Climate Change Resilience in Ten Asian Cities*, IDS Working Paper 315, Brighton, UK: Institute of Development Studies, http://onlinelibrary.wiley.com/store/10.1111/j.2040-0209.2009.00315_2.x/asset/j.2040-0209.2009.00315_2.x.pdf?v=1&t=izfwfhaa&s=690da7798426404dc6d4820ffbe1b20863a522f5 (accessed 21 February 2017).

UN/DESA (2012), *World Urbanization Prospects: The 2011 Revision*, http://esa.un.org/unpd/wup/pdf/WUP2011_Highlights.pdf.

UNISDR (United Nations Office for Disaster Risk Reduction) (2012), *Making Cities Resilient Report 2012*, www.unisdr.org.

VCAPS (2013), *Climate Adaptation Strategy: Ho Chi Minh City Moving towards the Sea with Climate Change Adaptation*, Ho Chi Minh City: VCAPS Consortium.

Webster, D. and P. McElwee (2009), 'Urban adaptation to climate change: Bangkok and Ho Chi Minh City as test beds', Paper presented at the 5th Urban Research Symposium, 5 June.

World Bank (2010), *Cities and Climate Change: An Urgent Agenda*, Washington, DC: World Bank.

WWF (2015), *Mega-Stress for Mega-Cities: A Climate Vulnerability Ranking for Major Coastal Cities in Asia*, Gland, Switzerland: WWF International.

12. Renewable energy economics

David Timmons

RENEWABLE ENERGY SOURCES

Every world region has some form of renewable energy, but availability and costs vary greatly across space and time. Renewable energy sources include biomass energy, hydropower, wind power, solar energy, wave energy, tidal energy, and geothermal energy. Many renewable sources produce energy in the form of electricity. A prominent feature of a renewable energy economy must thus be electrification, either directly (e.g. battery cars) or indirectly (e.g. cars powered by electrolytic hydrogen). In the following sections I refer mostly to electrical energy, though thermal energy from solar, biomass, and geothermal sources will likely contribute to renewable energy economies, and the principles described here apply equally to renewable thermal energy.

Biomass is any fuel derived from plant matter in the recent past, including wood, crops, crop residues, and animal waste. Biomass accounts for 10% of world primary energy supply and is the world's largest renewable energy source (IEA, 2014), since much of the world's population uses wood, charcoal, straw, or animal dung as cooking fuel. Biomass fuels also include wood chips, wood or grass pellets, ethanol, methanol, vegetable oil, and methane gas. The plant photosynthesis that creates biomass is only about 1% efficient, as compared to solar photovoltaic energy at perhaps 20% efficiency (Jacobson et al., 2015a). Collecting biomass energy thus requires large amounts of land, and biomass economics are to a large extent land economics.

Hydropower provides about 85% of world renewable electricity, and 16% of total electricity (IEA, 2012). Technical potential is about four times greater, equivalent to about 10% of current global energy use (IEA, 2014). Available energy is a function of water flow and head (or vertical fall), with the best sites having large values for both. Extent of hydropower development varies greatly by country. For example, Switzerland has developed 88% of its estimated technical potential, Norway 70%, China 24%, and the United States 16% (IEA, 2010).

For wind power, available energy is a function of the cube of wind velocity, so small velocity changes greatly affect power output and cost. The windiest sites tend to be coastal and offshore, on mountain ridges, and in large open plains. While wind turbines are visually prominent, their ground footprint is small compared to some other renewable energy sources.

Solar electricity may be generated from thermal energy in concentrating solar plants (CSP) or from photovoltaic (PV) cells. Here I focus on solar PV energy, the more widely usable form. Solar PV cells employ semiconductor material to generate a flow of electricity when struck by sunlight. In contrast to other renewable energy sources, solar PV is sustainably available in almost infinite quantities. In the United States, mean available solar PV energy varies from about 6.0 kWh/day/m^2 in New Mexico and Arizona to as little as 3.5 kWh/day/m^2 in coastal Oregon and Washington (NREL, 2008). Though these solar intensity differences affect costs, solar PV can be employed almost anywhere. Denholm and Margolis (2008) find that generating all U.S. electricity with solar PV would require an area about 235 km square, or 0.6% of U.S. land area, depending on panel location. If each state produced all its electricity from PV, the required land area ranges from 0.04% in Wyoming to 8.8% in New Jersey—about twice the current area of New Jersey roads and roofs.

For geothermal energy utilization, important variables include temperatures at different depths and ease of energy extraction. In geologically active areas, shallow wells produce steam for generating electricity in turbines, which unlike other renewable energy sources can run continuously. Binary cycle power plants operate at lower temperatures by boiling a nonwater liquid like ammonia, and are usable in more locations. More locations have geothermal water of sufficient temperature for space heating, and groundwater heat pumps use refrigeration technology to concentrate even lower temperature heat for thermal applications.

In this chapter I do not consider nuclear energy, since the characteristics of nuclear energy differ from those of the renewable energy sources discussed here. Nuclear economics hinge in part on the costs of improbable, infrequent, but extremely costly accidents, and on assumptions about nuclear waste costs, which may be incurred for millennia. The connection between peaceful and military uses of nuclear energy is another important issue, which economic tools have limited ability to evaluate.

RENEWABLE ENERGY COSTS AND COST FACTORS

Costs for energy sources are often expressed as levelized cost of energy (LCOE), or the cost per unit of energy based on amortized capital cost, assumed project life, present value of operating costs, and energy production. For example, in a solar PV system where for simplicity we assume no operating costs and no change in annual output, LCOE is the amortized cost divided by annual energy output:

$$LCOE = \frac{K\left(\frac{r(1+r)^T}{(1+r)^T - 1}\right)}{(P \cdot 8760 \cdot CF)} \tag{1}$$

where K is capital cost, r is an interest rate, T is the useful life of the project in years, P is the maximum power output, 8760 is the number of hours in a year, and CF is capacity factor. CF represents the portion of maximum annual energy production actually obtained, perhaps 0.2 in the case of solar PV (depending on location). Based on equation 1, greater CF clearly reduces LCOE.

Figure 12.1 compares the LCOE of different electricity sources. In some cases renewable energy appears more expensive than fossil-fuel alternatives. But avoiding the worst effect of climate change requires that fossil-fuel use be nearly eliminated (IPCC, 2014). Since the benefits of carbon neutrality are impossible to completely quantify, here I use a cost-effectiveness approach, where I accept carbon neutrality as a necessary goal and describe the least expensive way to meet this goal.

Note that in Figure 12.1 there is a considerable difference between the U.S. Energy Information Administration (EIA) and Lazard cost estimates, especially for the cost of solar energy. Such differences should narrow as the

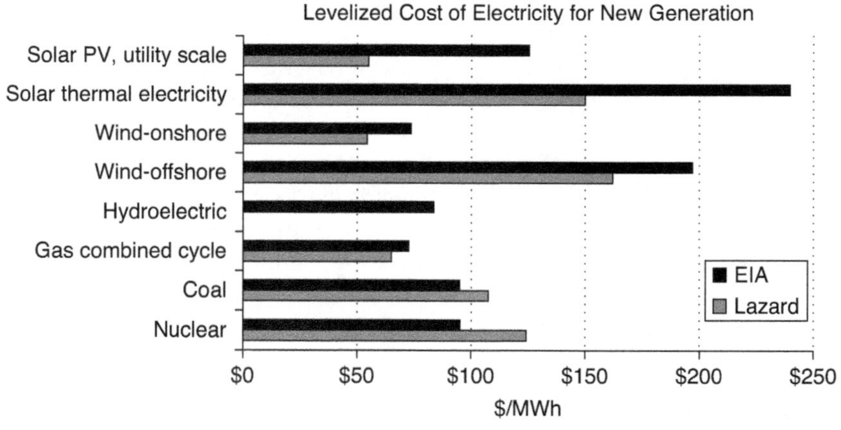

Sources: EIA (2015); Lazard (2015).

Figure 12.1 Cost comparison of renewable to fossil-fuel electricity

PV industry matures and the pace of technological change slows. All cost estimates are before any subsidies or tax benefits. The capital cost of U.S. residential and commercial solar energy fell from about \$12.00 per W_{DC} installed in 1998 to about \$4.25 per W_{DC} in 2013, a 65% reduction in 15 years (Barbose et al., 2014). While further cost decreases are likely, the extent of future declines is uncertain. For a complete comparison, the cost of renewable energy intermittency must also be included (as discussed below).

Renewable Energy Cost and Capital Intensity

Compared to fossil fuels, most renewable energy sources require large capital investments, as shown in Table 12.1. A large portion of fossil-fuel LCOE is for ongoing fuel purchases. While hydro, solar, wind, and geothermal energy sources have no fuel costs, significant capital expenditure is required.

The capital intensity of renewable energy means that renewable LCOE is sensitive to interest rates—changing interest rates effectively changes the cost of renewable energy. Unaffordable or unavailable capital may preclude construction of renewable energy facilities, and capital constraints could significantly affect the speed of transition to a renewable energy economy.

Table 12.1 Capital cost of renewable and non-renewable electricity sources

	Nominal capacity (MW)	Capital cost (million \$ per MW)	Assumed capacity factor	Capital cost (million \$ per expected MW*)
Natural gas: combined cycle	620	\$0.92	90%	\$1.02
Coal: advanced pulverized fuel	650	\$3.25	90%	\$3.61
Hydroelectric: conventional	500	\$2.94	75%	\$3.91
Nuclear: dual unit	2234	\$5.53	90%	\$6.14
Wind: onshore	100	\$2.21	25%	\$8.85
Biomass combined cycle	20	\$8.18	90%	\$9.09
Wind: offshore	400	\$6.23	35%	\$17.80
Solar: photovoltaic	150	\$3.87	20%	\$19.37
Solar: thermal electric	100	\$5.07	20%	\$25.34

Note: * For comparing sources with different capacity factors, we define million \$ per expected MW as (million \$/MW)/(capacity factor), or the capital cost to produce the same amount of electricity as 1 MW of capacity running continuously.

Source: Adapted from EIA (2013).

Site Sensitivity

The cost of renewable energy depends on where it is produced, in part because ambient energy varies greatly. Hydropower head varies by site, and flow varies continuously. Wind power varies with the cube of wind velocity; for example, doubling wind velocity results in eight times as much available energy. Solar PV in Seattle will produce only about half as much energy as in Phoenix, making Seattle solar electricity approximately twice as expensive. Ambient energy availability is a main driver of renewable energy cost differences. Capacity factor for wind power on a good site might be 33%, and for solar PV perhaps 20%. CF varies by type of renewable energy, and for any particular technology CF is generally indicative of ambient energy availability.

The scale of energy production can greatly affect cost. In 2015 installed cost per MW for 30 MW utility-scale solar PV projects was 43% of the cost of 5 kW residential-scale projects (Lazard, 2015) (note that 30 MW is 6000 times larger than 5 kW). In addition, energy production sites vary in terms of their ease of access, construction requirements, and proximity to the electricity grid.

All renewable energy sources require land, which varies greatly in cost. Biomass energy is particularly dependent on the cost of farmland and its productivity, since crop yields depend on soil characteristics, rainfall, and so on.

Even apparently similar locations can have differences in labor costs, regulatory procedures, engineering costs, and so on. One report urges "caution and specificity when referring to the installed price of PV, as clearly there is no single 'price' that characterizes the U.S. market, or even particular market segments, as a whole" (Barbose et al., 2014: 20).

RENEWABLE ENERGY COST MINIMIZATION

Marginal costs for renewable energy are typically rising in any particular place. If we build the first increments of hydropower, wind, solar, and geothermal power generation on the most accessible, least costly, and most productive sites, producing additional energy will be more expensive. For cost effectiveness, all energy sources within a given market should have the same marginal cost. For example, if new solar electricity were available for $100/MWh and wind electricity for $60/MWh, wind power would be the less expensive choice. But the next unit of wind energy would likely be more expensive (marginal cost rises). To minimize total energy cost, wind power should be chosen as long as it is less expensive than solar. If the

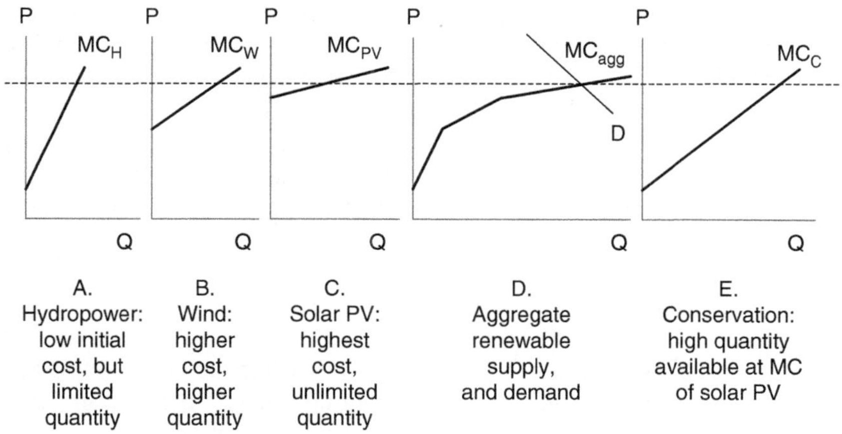

Figure 12.2 Equality of renewable energy marginal costs (MC) and marginal cost of conservation

marginal costs of energy sources differ, possibilities for substitution have not been fully exploited, and total cost is not minimized. The equimarginal principle states that minimum total cost occurs when all marginal costs are equal.

Figure 12.2 illustrates the equimarginal principle, with schematic representations of marginal cost (MC) curves illustrative of a renewable energy supply in some hypothetical region. Hydropower (panel A) is the least expensive renewable source for initial quantities, though costs rise quickly as more difficult hydropower sites are used. Wind (panel B) is somewhat more expensive but available in greater quantities than hydropower. Solar PV energy (panel C) has the greatest cost for initial quantities, but additional quantities near the same cost are virtually unlimited. Panel D shows an aggregate renewable energy marginal cost curve, a combination of curves A–C. Note that actual renewable energy supply functions will vary.

In Figure 12.2, panel D, the intersection of the aggregate supply curve and demand curve (D) identifies the quantity of energy produced in this hypothetical market. As in any market, the marginal cost of the most expensive resource needed to meet demand effectively determines the price of all resources used. Solar PV is the most expensive but most abundant renewable energy resource in this example, as in many parts of the world. In some regions offshore wind or geothermal energy might have a similar supply curve and thus play a similar role, setting the maximum price of all renewable energy.

The dotted horizontal line across panels A–D in Figure 12.2 shows that the marginal costs of all renewable energy sources used are equal. Though initial quantities of hydropower are less expensive than solar PV, the least-cost energy supply is provided by a combination of high-priced hydropower and wind, as well as solar (since in this example all three sources are needed to provide the required quantity of energy).

This discussion assumes a perfectly competitive market for energy (many buyers and sellers, perfect information, etc.), and no energy subsidies or taxes. While this is not typical of real energy markets today, these cost minimization principles should guide energy policy.

Energy Efficiency

Energy should be used more efficiently whenever this can be accomplished less expensively than producing more renewable energy: the marginal cost of energy efficiency should equal the marginal cost of renewable energy (Timmons et al., 2016). This is shown in Figure 12.2, panel E. Cost of conserved energy (CCE) in cost per MWh can be calculated with a formula similar to equation 1:

$$CCE = \frac{K\left(\frac{r(1+r)^T}{(1+r)^T - 1}\right)}{CE} \qquad (2)$$

where K is capital cost for a measure to improve energy efficiency, r is an interest rate, T is the useful life of the measure in years, and CE is the MWh of energy conserved as compared to a base case without the efficiency measure. Marginal costs of energy efficiency are often low compared to the cost of solar PV, the renewable that may determine the marginal cost of all renewable energy. Energy efficiency should thus play a major role in renewable energy economies. In non-industrialized countries with low energy use, this implies greater initial energy efficiency than in countries that have industrialized previously. For example, to minimize the cost of providing electric lighting, the first lighting technology used may be efficient light emitting diodes (LEDs) rather than the incandescent lighting used in the past.

Energy can be conserved both with technological changes, for example better building temperature controls, and with changing behaviors, for example increasing building temperatures in summer. Rising energy prices will create incentives for both types of conservation. Standards can also promote energy efficiency, such as standards for vehicle fuel use, buildings, appliances, electronics, and light bulbs. When efficiency standards are raised without increasing energy prices, there may be a leakage effect—lower operating cost may result in greater use, partly offsetting efficiency gains. This is one argument for accurately pricing all energy sources and for avoiding overreliance on energy standards and subsidies.

Marginal Cost of Renewable Energy: Temporal Considerations

Many renewable energy sources are unable to be delivered on demand, or are nondispatchable—we cannot control when the sun shines or the wind blows. This is a prominent issue in renewable energy economics, implying that the marginal costs of delivering renewable energy vary over time. For electrical energy, supply must always equal demand—unlike in other markets, temporary electricity shortages and surpluses can create severe problems. Approaches to intermittency include maintaining energy source diversity, pairing dispatchable and nondispatchable sources, and storing energy, all of which may increase renewable energy costs.

The equimarginal principle suggests that marginal costs should be equal for energy sources plus storage options in an energy system that can deliver energy at each moment in time. System design may be determined by critical conditions of peak demand and/or limited energy availability. For example, on a still night with no wind or solar energy availability the marginal cost of obtaining more hydropower should be equal to the marginal cost of wind or solar energy plus required storage.

Capacity factor (CF) is a key consideration in renewable energy system design. For simplicity, assume a constant 1 MW electricity demand. Without energy storage, completely utilizing a nondispatchable renewable energy source requires other energy for a 1 – CF portion of the supply. For example, a 1 MW wind turbine requires the 1 MW load to fully absorb peak wind output. With a CF of 0.33 and 8760 hours in a year, the wind turbines will supply (0.33×8760) MWh of electricity—this is the definition of CF. The electricity to be supplied by other sources is thus (0.67×8760) MWh. A diversity of intermittent sources, for example wind and solar, will supply a larger proportion of annual energy, but system costs may still be driven by occurrences of no wind or sun.

Intermittent sources may be paired with dispatchable sources, for example wind power with hydropower. Again assume a constant demand of 1 MW and 1 MW of wind capacity with a CF of 0.33, now paired with hydroelectric generation. Hydroelectric LCOE would be minimized with a constant output of 0.67 MW and CF equal to 1. But, in order to accommodate the intermittent wind production, hydroelectric CF must be 0.67, with a peak capacity of 1 MW and output ranging from 0 to 1 MW, depending on wind energy production. Greater costs are incurred by increasing the size of the hydropower turbine, generator, and penstock for

1 MW peak capacity instead of 0.67 MW. This cost of oversized hydropower is effectively a cost of wind power intermittency.

Energy storage is another alternative (Denholm et al., 2010). Continuing the same example, to fully supply the constant 1 MW load with wind power and energy storage, peak power must be 1/CF or (1/0.33) = 3 MW. Wind turbines with this capacity will fully supply the 1 MW electric load over the course of a year, given sufficient energy storage (for simplicity, ignoring energy losses in storage). Many energy storage technologies are available today (Beaudin et al., 2010; Díaz-González et al., 2012). Electrical storage options include: batteries and capacitors; pumped water storage, a hydropower variation; compressed air storage, for example in underground caverns; and energy stored as electrolytic hydrogen. Some of these technologies are also used to maintain power quality in grids with a high proportion of intermittent sources (Díaz-González et al., 2012). Storage of thermal energy figures prominently in some simulations of renewable energy systems (Palzer and Henning, 2014; Jacobson et al., 2015b). Intermittent production of energy-intensive products like nitrogen fertilizer is a way to store embodied energy.

Hydrogen storage may be especially important for energy resiliency, since hydrogen can be stored for long periods without energy loss, accommodating supply–demand variation at any time scale. Current limitations of hydrogen technology include the relatively high cost and low efficiency of both electrolysis and reconversion of hydrogen to electricity in a fuel cell. For example, a 2009 report estimates current electrolysis efficiency to be about 62% and fuel cell efficiency to be 47%, or just 29% overall storage efficiency (Steward et al., 2009).

Another approach is to use hydrogen gas directly as a fuel for transportation, heating, or industry, thus avoiding losses from fuel cell efficiency (but not losses in electrolysis). Oversized renewable energy generators can produce hydrogen when electricity supply exceeds demand (González et al., 2004; Jacobson and Delucchi, 2011). For example, a portion of perhaps 67% of the output of wind turbines with 2 MW capacity could directly supply a continuous 1 MW load (with the other portion of the 2 MW output exceeding the 1 MW demand, and the actual portion depending on the wind regime). The CF for direct use drops to (0.67 × 0.33) = 0.22. But 2 MW of turbine capacity still supplies more of the electric load than a 1 MW turbine (since 67% of 2 MW is greater than 100% of 1 MW), and 33% of the electricity generated could be used to produce hydrogen for fuel.

Using biomass energy is an approach to seasonal energy storage, since solar energy is chemically bound in biomass. Biomass technology is well developed, relatively low-cost, dispatchable on demand, and usable for either electricity or heat production. Biomass combined heat and power

(CHP) plants produce electricity and thermal energy simultaneously, and are more efficient than electric-only plants.

One study of renewable energy supply models PV production in the Southwest region of the United States, where capacity factors are higher and seasonal variation less than in other parts of the country, reducing costs of both energy production and storage (Mason et al., 2008; Fthenakis et al., 2009). Estimates include costs of interstate electricity transmission and compressed air energy storage (CAES), making solar PV energy a dispatchable electricity source comparable to fossil-fuel-powered electricity today. In this simulation the LCOE is $269/MWh, with the CAES component having a levelized cost of $39/MWh, or 14.5% of the total energy cost.

The marginal cost of accommodating intermittent renewable energy sources justifies greater cost for dispatchable renewable sources (since, as shown in Figure 12.2, at the minimum total cost all marginal costs are equal). Hydropower and geothermal sites that do not appear financially viable in today's fossil-fuel economy may be attractive for an energy system with all renewable sources.

To maximize social welfare, marginal energy cost should equal marginal benefit to consumers. If marginal costs vary over time, quantities demanded should also vary. This suggests using variable electricity pricing as another approach to intermittency. Consumers and businesses could forgo less critical activities at times of higher energy prices (DeCarolis and Keith, 2006).

Marginal Cost of Renewable Energy: Spatial Considerations

The variability of individual renewable energy sources is mitigated by having multiple sources in different geographic locations. The mean wind speed across an area is less variable than in any particular place. Increasing distance between generation sites typically reduces the covariance between sites, providing a more constant flow of energy (Hart et al., 2012).

Currently the United States does not rely greatly on interregional electricity flows, with less than 1% of 2010 delivered power moving between regional grids (EIA, 2011). In a renewable energy economy, long-distance energy movement may become more common. Capacity factors and associated marginal costs of renewable energy vary greatly by location, and high-voltage direct current (DC) transmission lines can move energy great distances with minimal energy loss (DeCarolis and Keith, 2006). In the Mason et al. (2008) study cited above, the estimated cost of interstate electricity transmission is $24/MWh, or 8.9% of total electricity cost, with the gains from greater CF in the Southwest greatly exceeding the cost of delivering energy to less sunny regions of the United States.

Long-distance electricity transmission costs are mostly fixed: if an investment is made in a transmission network, the marginal cost of moving electricity may be near zero (up to the design transmission capacity). This suggests that such long-distance electricity transmission facilities may be used not just for occasional backup, but for all energy production.

At the largest scales, the spatial boundaries of an energy system depend on both transmission costs and the limits of political and social organization. For example, a ring of linked solar energy production facilities situated around Earth's equator might provide energy for the entire world. The sun would shine continuously on some portion of this system, eliminating day–night energy storage, and there is no seasonal variation in day length at the equator. Yet in addition to the considerable engineering challenges, such a global energy system would require unprecedented international cooperation and interdependence.

The spatial equimarginal principle is that marginal costs of renewable energy from sources in different locations be equal, where marginal cost includes transmission from point of production to point of use. Depending on differences in production and transmission costs, this could imply that renewable energy costs are minimized with large, centralized production facilities, some of which would be at great distances from users. Yet such an energy system might not meet energy resiliency criteria. Optimal systems should balance energy cost and resiliency, where the precise marginal benefit of resiliency may be difficult to quantify. Just as many countries now subsidize domestic food production for greater food security, domestic and localized energy production may be attractive in a climate resilient world.

RENEWABLE ENERGY POLICIES AND CLIMATE RESILIENCE

At the beginning of this chapter I asserted that all energy will eventually be renewable—fossil fuels are finite, and available quantities can only decrease over time. Yet there is no certainty about the timing of a renewable energy transition, or about how much damage from climate change will occur in the meantime. There are three cost dynamics that affect the transition: rising fossil-fuel costs, declining renewable energy costs, and implementing policies that internalize fossil-fuel environmental costs.

Renewable energy will be adopted when fossil fuels are more expensive than renewables, that is, when fossil fuels are economically depleted. In Figure 12.3, increasing fossil-fuel extraction cost is indicated by the upward-sloping price path for fossil fuel (MC_{fossil}). Resource economists

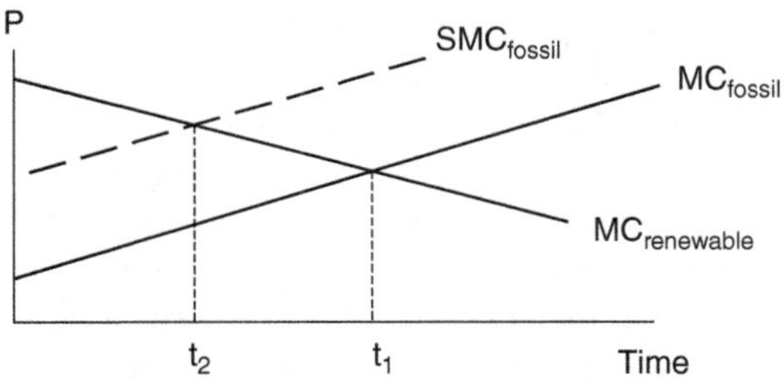

Figure 12.3 *Renewable energy transition dynamics*

typically expect rising marginal extraction costs based on decreased accessibility of fossil-fuel reserves (e.g. in deeper water, in shale formations), though historically fossil-fuel prices have sometimes fallen over time (Shafiee and Topal, 2010). At the same time as fossil-fuel prices rise, new technology will likely reduce renewable energy costs, as indicated by the downward-sloping price path for renewable energy ($MC_{renewable}$) in Figure 12.3. After time t_1 renewable energy will be less expensive than fossil fuel, and market forces may then accomplish a transition to renewable energy with little or no policy intervention.

Waiting until fossil fuels are economically exhausted could cause catastrophic damage from climate change. Internalizing fossil-fuel costs through carbon taxes or other policies is indicated by the social marginal cost path of fossil fuel in Figure 12.3 (SMC_{fossil}). Increasing the prices of fossil fuels with corrective taxes reduces the time of the transition to renewables from t_1 to t_2, and a large enough carbon tax could accomplish a renewable energy transition in the near future.

The transition to renewable energy also faces institutional and cultural obstacles. For example, energy supply diversity and intermittency may fundamentally change the role of electric utilities. And consumers may be slow to embrace required technologies such as electric vehicles. Policy approaches must include appropriate incentives and education.

SUMMARY

Many sources of renewable energy are available and have been used for centuries, and though costs have been greater than for fossil fuels in the past, cost differences are rapidly diminishing. Renewable energy costs depend in part on capital costs, site factors, and costs of managing intermittency.

Renewable energy cost is minimized when marginal costs for different renewable energy sources are equal in a system that provides energy at all times, and equal from all points in space. In a renewable energy economy, this may imply increasing use of dispatchable renewable energy sources including hydropower, geothermal, and biomass energy, as well as long-distance transmission of electricity to increase capacity factors and reduce energy storage requirements. Marginal cost of conserved energy should also equal marginal renewable energy cost, and the transition to renewable energy should be accompanied by increasing energy efficiency.

An eventual transition to renewable energy is inevitable, but the speed of the transition depends greatly on public policy choices. A combination of renewable energy sources, energy efficiency, energy storage, and energy transmission will eventually replace fossil fuels. Promoting climate change resilience suggests that this happen sooner rather than later.

REFERENCES

Barbose, G., S. Weaver and N. Darghouth (2014), *Tracking the Sun VII: An Historical Summary of the Installed Price of Photovoltaics in the United States from 1998 to 2013*, SunShot Initiative, U.S. Department of Energy and Lawrence Berkeley National Laboratory, http://emp.lbl.gov/publications/tracking-sun-vii-historical-summary-installed-price-photovoltaics-united-states-1998-20.

Beaudin, M., H. Zareipour, A. Schellenberglabe and W. Rosehart (2010), 'Energy storage for mitigating the variability of renewable electricity sources: An updated review', *Energy for Sustainable Development*, **14** (4), 302–314.

DeCarolis, J.F. and D.W. Keith (2006), 'The economics of large-scale wind power in a carbon constrained world', *Energy Policy*, **34** (4), 395–410.

Denholm, P. and R.M. Margolis (2008), 'Land-use requirements and the per-capita solar footprint for photovoltaic generation in the United States', *Energy Policy*, **36** (9), 3531–3543.

Denholm, P., E. Ela, B. Kirby and M. Milligan (2010), *The Role of Energy Storage with Renewable Electricity Generation*, Golden, CO: National Renewable Energy Laboratory.

Díaz-González, F., A. Sumper, O. Gomis-Bellmunt and R. Villafáfila-Robles (2012), 'A review of energy storage technologies for wind power applications', *Renewable and Sustainable Energy Reviews*, **16** (4), 2154–2171.

EIA (U.S. Energy Information Administration) (2011), *Today in Energy: Electricity Tends to Flow South in North America*, Washington, DC: EIA, https://www.eia.gov/todayinenergy/detail.cfm?id=4270.

EIA (U.S. Energy Information Administration) (2013), *Updated Capital Cost Estimates for Utility Scale Electricity Generating Plants*, Washington, DC: EIA, http://www.eia.gov/forecasts/capitalcost/.

EIA (U.S. Energy Information Administration) (2015), *Annual Energy Outlook 2015*, Washington, DC: EIA, http://www.eia.gov/forecasts/aeo/electricity_generation.cfm.

Fthenakis, V., J.E. Mason and K. Zweibel (2009), 'The technical, geographical, and economic feasibility for solar energy to supply the energy needs of the US', *Energy Policy*, **37** (2), 387–399.

González, A., E. McKeogh and B. Gallachoir (2004), 'The role of hydrogen in high wind energy penetration electricity systems: The Irish case', *Renewable Energy*, **29** (4), 471–489.
Hart, E.K., E.D. Stoutenburg and M.Z. Jacobson (2012), 'The potential of intermittent renewables to meet electric power demand: Current methods and emerging analytical techniques', *Proceedings of the IEEE*, **100** (2), 322–334.
IEA (International Energy Agency) (2010), *Renewable Energy Essentials: Hydropower*, Paris: IEA.
IEA (International Energy Agency) (2012), *Technology Roadmap: Hydropower*, Paris: IEA.
IEA (International Energy Agency) (2014), *Key World Energy Statistics*, Paris: IEA.
IPCC (2014), *Climate Change 2014: Synthesis Report, Summary for Policymakers*, Geneva: Intergovernmental Panel on Climate Change, https://www.ipcc.ch/pdf/assessment-report/ar5/syr/AR5_SYR_FINAL_SPM.pdf.
Jacobson, M.Z. and M.A. Delucchi (2011), 'Providing all global energy with wind, water, and solar power, part II: Reliability, system and transmission costs, and policies', *Energy Policy*, **39**, 1170–1190.
Jacobson, M.Z., M. Delucchi, G. Bazouin, Z.A. Bauer, C.C. Heavey, E. Fisher, S.B. Morris, D.J. Piekutowski et al. (2015a), '100% clean and renewable wind, water, and sunlight (WWS) all-sector energy roadmaps for the 50 United States', *Energy and Environmental Science*, **8**, 2093–2117.
Jacobson, M.Z., M.A. Delucchi, M.A. Cameron and B.A. Frew (2015b), 'Low-cost solution to the grid reliability problem with 100% penetration of intermittent wind, water, and solar for all purposes', *Proceedings of the National Academy of Sciences*, **112** (49), 15060–15065.
Lazard (2015), *Lazard's Levelized Cost of Energy Analysis – Version 9.0*, https://www.lazard.com/media/2390/lazards-levelized-cost-of-energy-analysis-90.pdf.
Mason, J., V. Fthenakis, K. Zweibel, T. Hansen and T. Nikolakakis (2008), 'Coupling PV and CAES power plants to transform intermittent PV electricity into a dispatchable electricity source', *Progress in Photovoltaics: Research and Applications*, **16** (8), 649–668.
NREL (National Renewable Energy Laboratory) (2008), 'Photovoltaic solar resource: Flat plate tilted south at latitude', http://www.nrel.gov/gis/images/map_pv_us_annual10km_dec2008.jpg.
Palzer, A. and H.-M. Henning (2014), 'A comprehensive model for the German electricity and heat sector in a future energy system with a dominant contribution from renewable energy technologies – part II: Results', *Renewable and Sustainable Energy Reviews*, **30**, 1019–1034.
Shafiee, S. and E. Topal (2010), 'A long-term view of worldwide fossil fuel prices', *Applied Energy*, **87** (3), 988–1000.
Steward, D., G. Saur, M. Penev and T. Ramsden (2009), *Lifecycle Cost Analysis of Hydrogen versus Other Technologies for Electrical Energy Storage*, Golden, CO: National Renewable Energy Laboratory (NREL).
Timmons, D., C. Konstantinidis, A.M. Shapiro and A. Wilson (2016), 'Decarbonizing residential building energy: A cost-effective approach', *Energy Policy*, **92**, 382–392.

PART III

Incentives, governance and policy

13. Carbon pricing policy design and revenue management: economic models and policy practice

Lint Barrage

INTRODUCTION

Over the past 25 years, a growing number of countries have adopted policies that place a price on the emissions of carbon dioxide (CO_2) and other greenhouse gases. Regardless of whether such carbon pricing is implemented through taxes or emissions trading schemes (ETS), these policies can raise substantial amounts of public funds. Carbon taxes raise revenues directly, and firms can be required to purchase emissions permits to comply with an ETS. How should the revenues raised by carbon pricing policies be managed? This chapter surveys both the economic models and the policy approaches that have been used to address this question from the perspectives of economic efficiency, equity, and political feasibility. First, public finance models typically find that the most efficient policy would create a revenue-neutral "tax swap" that (1) lowers tax rates that fall on capital income (e.g. corporate income taxes) and (2) uses the carbon revenues to make up for the lost revenues. Recycling the revenues through labor income tax cuts ranks second, whereas rebating the revenues directly to households as lump-sum payments entails the largest efficiency costs. In contrast, based on equity considerations, lump-sum rebates can be preferable to labor tax reductions in addressing the regressivity of carbon prices. In contrast, an ETS that gives emissions permits to firms for free and cuts corporate income taxes at the same time is likely the most regressive option, but would compensate industries and thus achieve political feasibility at the lowest cost.

In practice, the countries and regions that have successfully implemented carbon pricing policies to date have employed various combinations of these revenue recycling mechanisms. Indeed, many policies mandate a mix of revenue uses, such as rebates to low-income households along with income tax cuts for households and industry. Two features of observed policies stand out as indicative of the importance of political economy considerations. First, most ETS programs provide generous free permit allocations to firms. Second, many policies tie revenues directly to climate-related spending, such as clean energy programs. However, from an efficiency perspective, the funds for these programs should be raised through the least distortionary mix of broad-based tax instruments. Overall, however, the policy practice of carbon pricing revenue management has commonly struck a balance between efficiency, equity, and political considerations.

THEORY AND ECONOMIC MODELS

Efficiency

Carbon taxes

A rich literature has studied different carbon tax revenue recycling options from the perspective of economic efficiency. These studies typically compare the effects of using carbon pricing revenue to reduce different distortionary taxes (labor income taxes, corporate income taxes, etc.) or to be rebated as a lump sum.[1] A remarkable feature of these studies is the consistency of the efficiency rankings of different revenue recycling options across models. This is likely the case because the theoretical prediction for optimal policy is clear: carbon revenues should be used to decrease those taxes that impose the largest deadweight loss on the economy. That is, carbon pricing revenues should be used to offset the pre-existing taxes with the highest *marginal cost of public funds* (MCF).

The MCF measures the welfare cost of raising an additional dollar of government revenue from a given tax.[2] For example, Baylor and Beauséjour (2004) estimate that the MCF in Canada is 1.32 for personal income taxes and 1.37 for corporate income taxes. These results imply that the efficiency costs of raising another dollar in tax revenues are $0.32 for personal income taxes and $0.37 for corporate income taxes. In this setting, a marginal dollar of revenues should thus be used to reduce corporate rather than personal income taxes, as this would yield a higher efficiency gain. Importantly, it should be noted that the imposition of carbon prices is likely to change the MCF of other taxes in the economy. For example, by increasing energy prices, carbon taxes can increase the cost of consumption goods, thereby decreasing the real wage. If labor markets were already distorted by income taxes, then this additional decrease in the returns to labor can exacerbate the deadweight loss from labor taxes by further reducing labor supply. In other words, if carbon taxes decrease employment, they can thereby exacerbate the efficiency costs of pre-existing labor income taxes.[3] Analogous arguments apply to capital income and other taxes. Overall, the most efficient use of carbon pricing revenue is thus to reduce the tax rates with the highest marginal deadweight loss with the climate policy in place. That is, carbon pricing revenues should be used to counteract the efficiency costs they create.

Both theoretical models of optimal distortionary taxation (in the Ramsey tradition) and related computable general equilibrium (CGE) models often find that the most distortionary taxes are those that fall on capital income (e.g. corporate income taxes) or intermediate goods, or those taxes that have a narrow base. In contrast, the least distortionary taxes tend to be those falling on the consumption side of the economy, such as labor income taxes or consumption taxes, as well as taxes with a broad base (see, for example, Ballard et al., 1985; Goulder, 1995; Chari and Kehoe, 1998). Consequently, climate policy revenue recycling recommendations based on these models follow along similar lines, namely that capital income tax reductions should typically yield the highest efficiency gains, followed by labor and consumption taxes, and with lump-sum rebates as the least favorable option.[4] Intuitively, lump-sum transfers have an MCF of 1.00, as they are pure transfers between households and the government. There is thus no efficiency gain associated with returning climate policy revenues to households in lump-sum rebates. To the contrary, since these rebates increase households' incomes but do not increase the returns to working (i.e. real after-tax labor income), lump-sum rebates can create a pure income effect that reduces labor supply, thereby again exacerbating the welfare costs of labor income taxes. For this reason, they are generally found to be a less efficient revenue recycling option than tax cuts.

Table 13.1 summarizes the findings from several CGE models that have simulated the impacts of carbon taxes imposed under different revenue recycling scenarios. Indeed, these studies generally find the largest positive effects on GDP or welfare (equivalent variation, EV) from using carbon tax revenues to reduce capital income taxes, with labor taxes ranking second, and with lump-sum rebates as the costliest choice in terms of economic efficiency.

Emissions trading schemes

While the revenue-raising potential of carbon taxes is clear, the corresponding considerations in emissions trading schemes warrant further discussion. From a fiscal perspective, an emissions trading scheme can be designed to be formally equivalent to a carbon tax (abstracting from uncertainty considerations[5] as described by Weitzman (1974) and the subsequent literature). Consider a government seeking to reduce carbon

Table 13.1 *Carbon tax impacts across revenue recycling scenarios: economic modeling results*

Study	Initial carbon tax $/MT CO_2	Welfare/ outcome measure	Revenue recycling		
			Capital income	Labor income	Lump sum
Barrage (2014)[a]	$165 or $195 ($ 2005)[e]	EV $\%\Delta C_t \forall t$	+1.09%	+0.73%	
Carbone et al. (2013)[b]	$30 ($ 2012)	GDP (by 2035)	+1.00%	−0.5%	−3.5%
Goulder (1995)[c]	$92[e] ($ 1990)	Gross EV/$ revenue	−0.279	−0.374	−0.516
Jorgenson et al. (2013)[d]	$20 in 2020 ($ 2005)	Average EV/full wealth	+0.216	−0.080	−0.051
	$10 in 2020	Real GDP (2010–2050 average)	+0.2%	−0.1%	−1.0%
	$50 in 2020	Real GDP (2010–2050 average)	+0.4%	−0.7%	−3.7%
Jorgenson and Wilcoxen ([1996] 2002)	$55[e] in 1990 $238 in 2020	Real GNP (by 2020)	+1.10%	−0.69%	−1.70%

Notes:

a Average capital income tax cut is −9% (from 41.41% to 38.59%), and average labor income tax cut is −1.1% (from 36.58% to 36.17%).

b Tax cuts decrease marginal rates by same percentage-point amount in every bracket, equaling average change. Average capital tax cut −4.823%, average labor tax cut −1.413%, average lump-sum rebate $876.44 per adult per year.

c Marginal taxes on labor and capital cut −0.9% (for labor from 23.01% to 22.81%, for dividend interest from 22.89% to 22.72%, for accrual-equivalent tax on capital gains from 5.72% to 5.68%).

d Average tax rate cuts range from −5.2% ($10 tax) to −19.8% ($50 tax) for labor taxes, and −11.1% to −40.3% for capital income.

e Converted by the author from dollars per ton carbon (as reported in the study) into dollars per ton CO_2.

emissions from business-as-usual (BAU) level Q^{BAU} to Q^*. On the one hand, this reduction can be achieved by a carbon tax $\$\tau$ set equal to the aggregate marginal abatement cost MAC evaluated at Q^*. This tax will generate revenues worth $\$\tau \cdot Q^*$ paid by firms on their residual emissions. On the other hand, the government could impose an ETS restricting emissions to $\bar{Q}=Q^*$. Since firms will abate and trade permits until their marginal abatement costs are equated, the equilibrium permit price

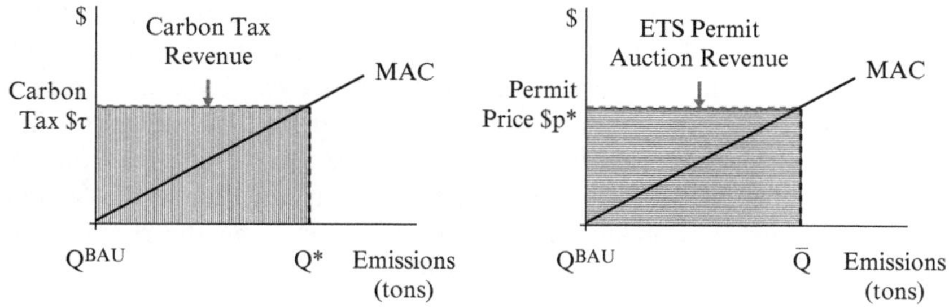

Figure 13.1 Revenue equivalence of carbon tax and ETS with 100 percent permit auctioning

will be $p^*=\$\tau=MAC(Q^*)$. Figure 13.1 illustrates the resulting revenue equivalence.

Importantly, an ETS can thus generate the same amount of public revenues as a carbon tax *if all emissions permits were auctioned off* to firms by the government. However, a much more common practice is for governments to give out the majority of permits for free. As discussed below, free permit allocations amount to a rent transfer to firms. In representative agent models – where the household owns the firms – this rent transfer can be shown to be formally equivalent to a lump-sum rebate of carbon pricing revenues to households. That is, an ETS with free permits is equivalent to a virtual emissions tax with lump-sum revenue rebates (see, for example, Goulder et al., 1999). Consequently, the efficiency costs of an ETS with non-auctioned permits correspond to those of lump-sum revenue recycling as outlined in Table 13.1. Conversely, the most efficient ETS would auction 100 percent of permits and use the revenues to cut taxes that fall on capital income (e.g. corporate income taxes).

Equity

Carbon taxes

One of the most important equity-based concerns about carbon taxes is that they are regressive. For example, Hassett et al. (2009) estimate that the burden of a carbon tax relative to income for households at the bottom decile of the income distribution would be between 1.25 and 5.3 times higher than the burden for the top decile, depending on the year and measure of income. However, these estimates do not consider the recycling of carbon tax revenues. A number of studies have shown that carbon pricing policies can be made *distributionally neutral* through appropriate revenue usage. Metcalf (2007) proposes an *environmental earned income tax credit* that would offset payroll taxes paid up to a limit. He studies the effects of a tax of \$15 per ton CO_2 in 2003 with a rebate capped at \$560 per worker. The cap makes the tax credit progressive, offsetting 73 percent of payroll tax payments for workers earning \$5000 per year but only 4 percent of payroll taxes for workers earning \$90 000 per year. In line with other studies, Metcalf (2007) estimates that the gross effects of a carbon tax would be regressive. However, the *net* effects of the tax after rebates would be close to distributionally neutral. Alternatively, using the revenues to provide households with lump-sum rebates of \$274 would make the policy fully progressive. Overall, Metcalf thus concludes that, while carbon taxes may be regressive in isolation, they can easily be designed to be distributionally neutral (or even progressive).

Studies exploring the distributional impacts of gasoline taxes have found similar results. West and Williams (2004a) estimate demand for gasoline, leisure, and other goods among different household types.[6] The authors then consider an increase in gasoline taxes to \$1.39 per gallon, the optimal overall tax as suggested by West and Williams (2004b). While they find that the tax is regressive in isolation, its net incidence depends on how the revenues are used. On the one hand, the tax could be made essentially progressive by rebating revenues to consumers in a lump-sum fashion. Indeed, the transfers more than offset the gasoline cost increase for the bottom quintile of the income distribution. However, this policy entails the largest efficiency costs. In contrast, using revenues to reduce labor income taxes – specifically through an equal percentage-point cut in all income tax brackets – mitigates the regressivity and reduces the efficiency costs (relative to lump-sum rebates), thus arguably presenting an attractive potential compromise.

Bento et al. (2009) estimate a multi-market model of U.S. vehicle and driving demand and supply to study the effects of a \$0.25 per gallon gasoline tax increase. They consider three revenue recycling scenarios: lump-sum rebates, income-based (proportional to households' income shares), and proportional to households' baseline share of vehicle miles traveled (VMT). It should be noted that the model is not designed to consider revenue recycling through reductions in distortionary taxes, such as labor income taxes. Consequently, the authors can only consider different types of lump-sum transfers. However, the computed gasoline tax welfare impacts include general equilibrium effects such as changes in vehicle prices and carmaker profits. Despite the difference in the model and study focus, the authors also find that standard lump-sum revenue recycling more than offsets the regressivity of the gasoline tax. Lower-income groups are made better off, as they receive a larger share of tax revenues than their share of gasoline tax payments. With income-based rebates, the welfare effects of the tax are U-shaped when plotted across income deciles. Intuitively, this is because medium-income households have the highest ratio of miles driven – and thus gasoline taxes paid – relative to income. An income-based rebate thus compensates them relatively less than households on either extreme of the income distribution. Finally, VMT-based rebates render the gasoline tax increase close to distributionally neutral, although higher-income households are made slightly worse off. Overall, however, the study's results thus imply once again that lump-sum rebates can be used to mitigate or even reverse the regressivity of gasoline taxes.

Finally, prior work on carbon pricing revenues in the electricity sector has found broadly similar results, but raised its own set of complications and caveats. For example, Burtraw et al. (2009b) compare the use of emissions trading permit auction revenues to provide free allocations to electricity and natural gas local distribution companies with rebating the value of this effective subsidy to households directly. With the exception of the top income decile, most households are better off with the lump-sum rebate than with the electricity subsidy. However, the distribution of welfare impacts depends on nuances such as whether utilities pass through the free permit values through decreases in variable or fixed electricity charges, as well as on whether households respond to such price changes differently (rationally).

Emissions trading schemes

In theory, carbon emissions trading schemes could raise the same amount of revenues as carbon taxes by auctioning off emissions permits (see Figure 13.1). In this case, the distributional effects of competing policy options are in line with those predicted for carbon taxes. For example, Burtraw et al. (2009a) model five revenue recycling options for a U.S. cap and trade program. They find that the policy can be made progressive if revenues are rebated to households in a lump sum, or through an expansion of the earned income tax credit. In contrast, the policy remains regressive if permit auction revenues are used to reduce income or payroll taxes.

In reality, at least some emissions permits are commonly given out for free. This policy has two main implications for the distributional effects of an ETS. First, the forgoing of revenues limits a government's ability to implement tax and rebate policies to mitigate the regressive effects of energy price increases associated with the ETS. Second, the benefits of free permits may accrue to shareholders of firms receiving these rents. The extent to which equities are more likely to be owned by wealthier households may thus further exacerbate the regressivity of an ETS. Several studies have quantified these effects.

Dinan and Rogers (2002) study the distributional effects of an ETS designed to reduce U.S. carbon emissions by 15 percent. They compare two permit allocation schemes: free distribution (where the government re-captures 45 percent of the resulting rents as "residual revenue" through income and profit taxation) and full auctioning (where the government captures 100 percent of the rents, as with a carbon tax). The authors also consider three competing revenue recycling schemes: corporate tax cuts, payroll tax cuts, and lump-sum rebates. The results suggest that the gross costs of the climate policy are regressive. In line with other studies, full auctioning of permits with lump-sum rebates could eliminate this regressivity. However, all other policies considered remain regressive. In particular, free permit allocation renders the ETS regressive even with lump-sum rebates of the residual revenues raised through income and profit taxes, costing households in the bottom quintile of the income distribution an estimated 0.4 to 4.2 percent of annual income, compared with a cost of −0.5 to 0.0 percent for households in the top quintile. The regressivity is most severe with free permits and residual revenues used to reduce corporate income taxes, costing bottom quintile households an estimated 2.4 to 6.2 percent of annual income, compared with −0.6 to −1.2 percent for top quintile households.

Parry (2004) finds similar results in studying an ETS that would reduce U.S. carbon emissions from the electricity sector by 10 percent. He estimates that the gross burden of the policy would amount to \$106 for households in the bottom quintile of the income distribution and \$406 for households in the top quintile. However, the profit income resulting from free permit allocations and the associated rents would yield \$370 for top quintile households but only \$24 for bottom quintile households. The reason for this discrepancy is that, based on the Consumer Expenditure Survey (CEX), he estimates that top income quintile households own 53 percent of stockholdings (within the CEX sample), compared with only 3.5 percent for lowest income quintile households. Parry also considers that the government can re-capture 35 percent of the permit rents through income and profit taxation. However, even after lump-sum distribution of these revenues, the ETS remains regressive: the net burden amounts to 0.061 percent of income for households in the bottom quintile, compared with a −0.063 percent burden (or net gain) for households in the top quintile. In contrast, a carbon tax (or an ETS with 100 percent permit auctioning) with lump-sum revenue recycling would be progressive, yielding an estimated net cost of −\$109 for households in the bottom quintile, compared with a cost of \$191 for the top quintile.

Metcalf (2007) also considers an ETS with free permits in his analysis. He finds that such a policy would be "decidedly regressive," decreasing households' disposable income the most for the lowest income deciles, and increasing income for only the two top deciles.

Overall, free permit allocation in an ETS thus appears to greatly increase the regressivity of the policy across income groups, even if some rents are re-captured through income taxation and distributed to households in lump-sum payments. Unfortunately, this finding creates an important dilemma for policy makers, as such partially free permit allocations and rent transfers to shareholders are precisely what would be needed to render an ETS distributionally neutral from the perspective of affected industries, as described in the next subsection.

Political Feasibility

Lack of political feasibility is one of the largest obstacles to the enactment of climate policies. In turn, one of the key determinants of the political feasibility of different policies is the distribution of their impacts across industries (Bovenberg and Goulder, 2001; Bovenberg et al., 2008). Policy revenues can be used to modify these impacts in several ways, including corporate tax cuts, transfers, and the giving out of free permits in the context of an emissions trading scheme. As discussed above, the distribution of free permits constitutes a rent transfer to firms. Intuitively, the creation of an ETS effectively restricts output, causing prices to rise. However, with free permits, firms do not face an associated increase in production costs, thus earning rents through the free permit allocation (Bovenberg and Goulder, 2001). Despite the efficiency costs of this policy (see, for example, Goulder et al., 1999), its potential value in enhancing the political feasibility of an ETS is evident.

Bovenberg and Goulder (2001) employ a dynamic computable general equilibrium model of the U.S. economy to study the effects of a carbon pricing policy on output, after-tax profits, and equity values of firms under competing revenue recycling and industry compensation schemes. In particular, they consider policy adjustments that would protect regulated industries from adverse effects on their equity values or overall tax bills. More formally, they consider policies that achieve either "equity-value neutrality" – implying that the real value of the most vulnerable industries remains unchanged by the announcement and enactment of the climate policy – or "tax-payment neutrality," ensuring that an industry's overall tax payments (from carbon taxes, corporate taxes, indirect labor taxes, etc.) remain unchanged. These distributional goals are achieved through (1) industry-specific corporate tax cuts, (2) lump-sum transfers to industry-specific capital, or (3) inframarginal carbon tax exemptions, which are equivalent to an ETS with some free permit distribution.

The study's most significant finding is that compensation for the most vulnerable industries could be achieved at little to no efficiency cost. Consider, for example, a carbon pricing policy that recycles all revenues through personal income tax (PIT) reductions, against a policy that first ensures equity-value neutrality through industry-specific corporate income tax adjustments, and then uses the remainder of the revenue for PIT reductions. Perhaps surprisingly, the policy that achieves equity-value neutrality is *more efficient* than the standard policy. That is, the gross efficiency costs are \$60.00 per ton CO_2 emissions with the standard policy, but only \$46.90 per ton with vulnerable industry compensation through targeted corporate income tax reductions. Intuitively, this is because corporate income taxes are more distortionary than personal income taxes, and consequently this effective tax swap improves overall efficiency. Similarly, a policy using lump-sum payments to affected industries achieves equity-value neutrality with only a minor gross efficiency cost increase (\$61.40 per ton CO_2) compared to the all-PIT reduction policy (\$60.00 per ton). Finally, free permits appear to be an effective and low-cost way of achieving some distributional neutrality. The estimated share of emissions permits that must be given out for free to achieve equity-value neutrality is only 4.3 percent in the coal industry, and 15 percent for oil and gas. The associated gross efficiency costs of \$64.60 per ton CO_2 compare favorably with those of the standard policy raising and using 100 percent of revenues to reduce personal income taxes (\$60.00 per ton).

It is important to note that more generous free permit allocation schemes – such as those commonly observed in practice – create significantly higher efficiency costs. For example, a policy granting firms 90 percent of their base year business-as-usual emissions as free ("grandfathered") permits each year increases the gross efficiency costs to \$77.40 per ton CO_2. Such policies are also predicted to *increase* equity values in the most vulnerable industries, notably coal and oil and gas, thus arguably overshooting the goal of industry compensation.

The authors also consider industry compensation to achieve tax-payment neutrality. Importantly, the tax cuts or transfers required to achieve such neutrality are considerably larger than those needed for equity-value neutrality. Intuitively, this is because supply in the most vulnerable industries is quite elastic, implying that energy sector firms can pass on most of their tax burdens to consumers. Consequently, a tax-payment neutrality policy overcompensates firms relative to the true economic burden of the climate policy. However, the efficiency costs of even this generous compensation policy compare favorably (\$61.10 per ton CO_2) with those of a fully PIT revenue recycling scheme as long as tax-payment neutrality is achieved through targeted corporate income tax reductions.

Overall, the key finding of Bovenberg and Goulder's (2001) study is thus that climate policies can be designed to protect profits in the most vulnerable industries at minimal additional efficiency costs. The most efficient policy would use industry-specific corporate income tax cuts to achieve equity-value efficiency; however, distributing modest amounts of free permits to key industries in an ETS can achieve the same goal. While political feasibility adjustments remain an important area for future research, some subsequent studies have found similar results. Bovenberg et al. (2005) find that, while the share of permits that must be freely allocated to compensate firms is increasing in the emissions reductions target of the policy, this share remains below 50 percent for a wide range of targets and parameters. Similarly, Burtraw et al. (2002) find that it would cost only 7.5 percent of the value of auctioned emissions allowances to compensate electricity generators for the losses to their asset values due to the imposition of an ETS in the electricity sector (see also Burtraw and Palmer, 2008).

Theory Summary

Many economic models of fiscal policy and carbon pricing produce similar conclusions on the rankings of different revenue usage schemes in order to achieve a given policy objective. However, *across* policy objectives, the rankings of preferred policies differ starkly. Policy makers consequently face a trade-off between efficiency, equity, and political feasibility.

For example, from the perspective of economic efficiency, the preferred policy would be a carbon tax (or an ETS with 100 percent auctioned permits) with revenues recycled fully in cuts on taxes that fall on capital income, such as corporate income taxes. Rebating revenues as lump-sum payments to households would yield the lowest efficiency gains. In contrast, from the perspective of equity, such lump-sum rebates would be the best policy. The least equitable or most regressive policy would be an ETS with free permits and residual revenue recycled through corporate income tax cuts. However, partially free permits and targeted corporate income tax cuts would be the least costly choice to ensure industry compensation, and thus political feasibility.[7] While economic models can thus provide guidance to policy makers, they generally do not yield a unique "best" revenue recycling method that would hold across competing policy objectives.

POLICY PRACTICE

Over the past 25 years, a growing number of countries and regions across the world have adopted carbon pricing policies (for details, see, for example, Sumner et al., 2009; World Bank, 2014). The policy practice with regard to revenue usage varies greatly across jurisdictions, and often differs markedly from the recommendations of economic models. Table 13.2 summarizes revenue usage for a select sample of policies.

Table 13.2 Carbon pricing revenue usage in policy practice

Location	Carbon policy	Year	Percentage of carbon pricing revenues* allocated to:								
			Tax rate reductions			Lump-sum transfers		Program expense		General government fund	Free permits
			Income		Payroll	Households	Industry	Green technology	Low income energy		
			Personal	Corporate							
Australia	Tax/ETS	2013/2014[a]	✓(26%)*			✓(25%, targeted)	✓(1.7%, targeted[b])	✓(14%)	✓(1.5%[c])	✓	✓(48%)
British Columbia	Tax	2013/2014	✓(20%)	✓(36%)		✓(22%, targeted)					
Denmark	Tax	2000[d]			✓(82%)			✓(19%)			
France	Tax	2016[e]			✓(75%)	✓(targeted)		✓			
Iceland	Tax	2010								✓(100%)	
Ireland	Tax	2010						✓(15%[f])		✓(100%)	
Japan	Tax	2012						✓(100%)			
Norway	Tax	2007								✓(100%)	
Portugal	Tax	2015	✓(100%)								
Quebec	ETS	2015						✓			✓
Sweden	Tax	2009								✓(100%)	
Switzerland	Tax	2015				✓(41%, general)	✓(25%, general)	✓(33%)			
RGGI[g]	ETS	2015	Member states have authority over permit auction revenues. Majority of aggregate funds spent on energy efficiency, clean and renewable energy, household energy bill assistance, etc.								✓(31%)
EU	ETS	2013	Member states have authority over permit auction revenues. EU Directive requests that "at least 50%" be used for climate and energy-related purposes.								✓(60%)

Notes:

* Percentage indicates approximate share of carbon pricing revenues in indicated year allocated to a given purpose. Shares may not add to 100 as a result of other spending or shortfalls.

a Calculations based on projections from Australian Government's 2011 "Clean Future Final Plan" as summarized by Carl and Fedor (2012). Expenditures may exceed or fall short of revenue raised. For 2013/2014, more revenues were allocated than projected to be raised through the carbon pricing scheme (Carl and Fedor, 2012).

b Figure excludes government buyouts of inefficient coal-fired power plants, as their value is included in coal industry assistance through free permits.

c Includes all *household* energy efficiency expenditures.

d Percentage of total revenue from taxes on both CO_2 (€277 million) and SO_2 (€42 million). Note that total expenditures included some additional categories (e.g. funds for self-employed) and exceeded revenues slightly in 2000 after adjustments (Danish Energy Authority).

e *Projected* revenue usage, with details on spending besides competitiveness (payroll) tax credit (CICE) still to be determined (Elbeze, 2014).

f While €50 million from the estimated carbon tax revenues of €330 million were designated for energy efficiency programs (Irish Government, 2009), these funds are not generally earmarked as such (ICMM, 2013).

g Regional Greenhouse Gas Initiative, an ETS of the electricity sector across a group of northeastern U.S. states.

Sources: BAFU; European Commission; Government of Iceland; Government of Ireland, Annexes to Summary of 2010 Budget Measure; International Energy Agency; Parliament of Australia; Sumner et al. (2009); Carl and Fedor (2012); ICMM (2013); Governo de Portugal (2014); Government of British Columbia (2015); Pereira and Rodrigues (2015); Québec Ministère du Développement durable, de l'Environnement et de la Lutte contre les changements climatiques (2015); RGGI (2015a, 2015b).

Three main insights and surprises emerge from these figures. First, the majority of policies use revenues for expenditure programs such as for clean and renewable energy technology, energy efficiency, and the environment. As noted above, economic models often exclude this revenue usage option. On the one hand, the calibration of government spending effects on the macroeconomy and/or directed technical change is highly controversial. On the other hand, standard public finance theory predicts that the best way to raise revenues for any program is through the least distortionary tax instrument. Depending on the level of carbon taxes, it may thus be preferable to raise the next dollar of revenue for clean energy subsidies from labor or consumption taxes. Intuitively, carbon taxes generally fall on intermediate inputs (energy) and are thus an undesirable tax base above and beyond externality levies (see, for example, Goulder, 1995; also discussion in Barrage, 2014). In theory, clean energy subsidies should thus be set based on efficiency criteria (e.g. to correct market failures in R&D incentives) and financed out of general government funds. In reality, however, owing to political economy frictions, it may be desirable to earmark some carbon pricing revenues for clean energy subsidies (see, for example, Brett and Keen, 2000). In particular, if political economy frictions prevent an optimized provision of public goods from general funds, then it may be necessary to tie carbon pricing revenues to expenditures on clean energy research or other environmental programs in order to expand the provision of these goods.[8]

The second surprise is that, with the exception of British Columbia, none of the policies considered recycled carbon pricing revenues through capital income tax reductions, despite the consistent prediction from economic models that this would be the most efficient revenue usage.

Third, the main finding is that many countries appear to balance competing policy objectives. Most countries appear to split revenues across multiple uses so as to at least partly protect the interests of vulnerable households, industry, and the broader economy.

Many more countries have introduced or are currently introducing carbon pricing policies, including large, emerging market economies such as China, Mexico, Chile, and South Africa. These carbon pricing schemes will undoubtedly play a critical role in building global climate resilience. The landmark Paris Agreement on climate change – reached in December 2015 at the 21st meeting of the Conference of the Parties to the United Nations Framework Convention on Climate Change – saw many countries pledging emission reduction targets supported by carbon pricing policies. Importantly, the world's largest carbon emitter – China – has already implemented pilot emissions trading schemes in several cities, and pledged its intent to build a nationwide ETS. Understanding the impacts of competing carbon pricing design elements will thus remain an important and policy-relevant area for future research in economics and climate resilience.

ACKNOWLEDGMENTS

I thank Amy Ng for excellent research assistance on this project. I also thank K.S. Kavi Kumar, David Simpson, and Karachepone Ninan for their detailed and helpful feedback on this chapter.

NOTES

1. In contrast, changes in the provision of public goods – such as using revenues to fund new infrastructure projects – are often not considered. On the one hand, the quantification of the economic benefits of increases in public spending are highly controversial. On the other hand, formally endogenizing government spending in the literature's workhorse dynamic taxation (Ramsey) models complicates the models considerably.
2. Closely related concepts include the marginal excess burden, the marginal deadweight loss, and the marginal welfare cost of taxes. Though similar, the formal definition of these concepts can differ (see, for example, Triest, 1990; Fullerton, 1991; Snow and Warren, 1996; and Dahlby, 2008).
3. This "tax interaction effect" is typically found to exceed the efficiency gains from optimized carbon tax revenue recycling (see, for example, Goulder, 1995; Parry, 1995; Bovenberg and Goulder, 1996). In these and other standard Ramsey taxation models, a "double dividend" does therefore not arise without additional model features, such as untaxable rents (Bento and Jacobsen, 2007).
4. While tax codes differ greatly across countries, the general finding that taxes falling on capital income (e.g. corporate income taxes) entail higher efficiency costs than those falling on labor emerges from studies estimating the MCF across several countries. See, for example, the international MCF estimate review table in the Online Appendix of Barrage (2014) located at http://lintbarrage.com/wp-content/uploads/2014/12/Barrage_2014_OnlineAppendix.pdf.
5. It should be noted that uncertainty is an important research frontier, both for economic analyses of climate policy in general (see, for example, the review by Burke et al., 2016) and for the study of fiscal–climate policy interactions in particular, where uncertainty has been largely abstracted from.
6. The authors estimate an Almost Ideal Demand System from Consumer Expenditure Survey data in order to account for households' (heterogeneous) demand responses to potential gasoline tax increases.
7. It should be noted that the discussion in this chapter and cited studies focuses on political feasibility in terms of industry asset values and profitability. In reality, feasibility is likely intertwined with other policy aspects, including its distributional impacts. Importantly, political feasibility may further depend on distributional effects along dimensions besides income, such as age or geography. I thank David Simpson for the insight that the delineation of "efficiency," "distributional," and "political feasibility" objectives is thus "somewhat fluid and inexact."
8. I thank Kavi Kumar for this insight.

REFERENCES

BAFU (Swiss Federal Bundesamt für Umwelt), *Rückverteilung der CO2-Abgabe*, Bern: BAFU, http://www.bafu.admin.ch/klima/13877/14510/14749/index.html?lang=de (accessed November 2015).

Ballard, Charles L., John B. Shoven and John Whalley (1985), 'General equilibrium computations of the marginal welfare costs of taxes in the United States', *American Economic Review*, **75** (1), 128–138.

Barrage, Lint (2014), 'Optimal dynamic carbon taxes in a climate-economy model with distortionary fiscal policy', Working Paper, http://www.lintbarrage.com/wp-content/uploads/2016/12/Barrage-2016.pdf.

Baylor, Max and Louis Beauséjour (2004), *Taxation and Economic Efficiency: Results from a Canadian CGE Model*, Ottawa: Department of Finance Canada.

Bento, Antonio M. and Mark Jacobsen (2007), 'Ricardian rents, environmental policy and the "double-dividend" hypothesis', *Journal of Environmental Economics and Management*, **53** (1), 17–31.

Bento, Antonio M., Lawrence H. Goulder, Mark R. Jacobsen and Roger H. van Haefen (2009), 'Distributional and efficiency impacts of increased US gasoline taxes', *American Economic Review*, **99** (3), 667–699.

Bovenberg, A. Lans and Lawrence H. Goulder (1996), 'Optimal environmental taxation in the presence of other taxes: General-equilibrium analyses', *American Economic Review*, **86** (4), 985–1000.

Bovenberg, A. Lans and Lawrence H. Goulder (2001), 'Neutralizing the adverse industry impacts of CO_2 abatement policies: What does it cost?', in C. Carraro and G. Metcalf (eds), *Behavioral and Distributional Effects of Environmental Policy*, Chicago, IL: University of Chicago Press, pp. 45–85.

Bovenberg, A. Lans, Lawrence H. Goulder and Derek J. Gurney (2005), 'Efficiency costs of meeting industry-distributional constraints under environmental permits and taxes', *RAND Journal of Economics*, **36** (4), 951–971.

Bovenberg, A. Lans, Lawrence H. Goulder and Mark R. Jacobsen (2008), 'Costs of alternative environmental policy instruments in the presence of industry compensation requirements', *Journal of Public Economics*, **92** (5), 1236–1253.

Brett, Craig and Michael Keen (2000), 'Political uncertainty and the earmarking of environmental taxes', *Journal of Public Economics*, **75** (3), 315–340.

Burke, M., M. Craxton, C.D. Kolstad, C. Onda, H. Allcott, E. Baker, L. Barrage, R. Carson et al. (2016), 'Opportunities for advances in climate change economics', *Science*, **352** (6283) (15 April), 292–293.

Burtraw, Dallas and Karen Palmer (2008), 'Compensation rules for climate policy in the electricity sector', *Journal of Policy Analysis and Management*, **27** (4), 819–847.

Burtraw, Dallas, Karen Palmer, Ranjit Bharvirkar and Anthony Paul (2002), 'The effect on asset values of the allocation of carbon dioxide emission allowances', *Electricity Journal*, **15**, 51–62.

Burtraw, Dallas, Richard Sweeney and Margaret Walls (2009a), *The Incidence of U.S. Climate Policy*, RFF Discussion Paper 0 9–17, Washington, DC: Resources for the Future.

Burtraw, Dallas, Margaret Walls and Joshua Blonz (2009b), *Distributional Impacts of Carbon Pricing Policies in the Electricity Sector*, RFF Discussion Paper 09–43, Washington, DC: Resources for the Future.

Carbone, Jared, Richard D. Morgenstern, Roberton C. Williams III and Dallas Burtraw (2013), *Deficit Reduction and Carbon Taxes: Budgetary, Economic, and Distributional Impacts*, Report, Washington, DC: Resources for the Future.
Carl, Jeremy and David Fedor (2012), 'Revenue-neutral carbon taxes in the real world: Insights from British Columbia and Australia', Hoover Institution, Stanford University.
Chari, V.V. and Patrick J. Kehoe (1998), 'Optimal fiscal and monetary policy', Federal Reserve Bank of Minneapolis Staff Report 251.
Dahlby, Bev (2008), *The Marginal Cost of Public Funds*, Cambridge, MA: MIT Press.
Danish Energy Authority, 'Green taxes in trade and industry – Danish experiences', http://www.ens.dk/sites/ens.dk/files/consumption-savings/energy-consumption-production-industries/Green_taxes%20danish%20experiences.pdf (accessed November 2015).
Dinan, Terry and Diane Lim Rogers (2002), 'Distributional effects of carbon allowance trading: How government decisions determine winners and losers', *National Tax Journal*, **55** (2), 199–221.
Elbeze, Jeremy (2014), 'The reform of energy taxation: An extension of carbon pricing in France', Conference presentation, http://conferences.au.dk/fileadmin/conferences/gcet/Presentations_in_the_detailed_programme/T_125_14-09-26_Presentation_French_Reform_of_energy_Taxation_Jeremy_ELBEZE.pdf.
European Commission, http://ec.europa.eu/clima/policies/ets/cap/auctioning/index_en.htm (accessed November 2015).
Fullerton, Don (1991), 'Reconciling recent estimates of the marginal welfare cost of taxation', *American Economic Review*, **81** (1), 302–308.
Goulder, Lawrence (1995), 'Effects of carbon taxes in an economy with prior tax distortions: An intertemporal general equilibrium analysis', *Journal of Environmental Economics and Management*, **29**, 271–297.
Goulder, Lawrence, Ian W.H. Parry, Roberton C. Williams III and Dallas Burtraw (1999), 'The cost-effectiveness of alternative instruments for environmental protection in a second-best setting', *Journal of Public Economics*, **72** (3), 329–360.
Government of British Columbia (2015), *Budget and Fiscal Plan 2015/16–2017/18*, http://bcbudget.gov.bc.ca/2015/bfp/2015_budget_and_fiscal_plan.pdf (accessed November 2015).
Government of Iceland, 'The Fiscal Consolidation Plan 2011', https://eng.fjarmalaraduneyti.is/media/Fjarlog/Fiscal_consolidation_plan_2011.pdf (accessed February 2017).
Government of Ireland, Annexes to Summary of 2010 Budget Measure, http://budget.gov.ie/budgets/2010/Documents/Annexes%20to%20the%20Summary%20of%20Budget%20Measures%20Final.pdf (accessed February 2017).
Governo de Portugal (2014), *The Green Growth Commitment and the Green Taxation Reform*, http://www.crescimentoverde.gov.pt/wp-content/uploads/2014/10/2014_12_05_Portugal-green-taxation-reforma-and-green-growth-deal.pdf (accessed November 2015).
Hassett, Kevin A., Aparna Mathur and Gilbert E. Metcalf (2009), 'The incidence of a US carbon tax: A lifetime and regional analysis', *Energy Journal*, **30** (2), 155–177.
ICMM (International Council on Mining and Metals) (2013), *Options in Recycling Revenues Generated through Carbon Pricing*, Report, London: ICMM.
International Energy Agency, http://www.iea.org/policiesandmeasures/pams/europeancommission/name-24558-en.php (accessed November 2015).

Irish Government (2009), 'Financial statement of the Minister for Finance Mr. Brian Lenihan, T.D., 9 December 2009', http://www.budget.gov.ie/Budgets/2010/FinancialStatement.aspx#item9 (accessed February 2017).

Jorgenson, Dale W. and Peter J. Wilcoxen ([1996] 2002), 'Results from the Jorgenson–Wilcoxen model reducing US carbon emissions: An econometric general equilibrium assessment', in Lawrence H. Goulder (ed.), *Environmental Policy Making in Economies with Prior Tax Distortions*, Cheltenham, UK and Northampton, MA, USA: Edward Elgar Publishing, pp. 276–292.

Jorgenson, Dale W., Richard J. Goettle, Mun S. Ho and Peter J. Wilcoxen (2013), *Double Dividend: Environmental Taxes and Fiscal Reform in the United States*, Cambridge, MA: MIT Press.

Metcalf, Gilbert E. (2007), *A Proposal for a U.S. Carbon Tax Swap*, Discussion Paper 2007–12, Hamilton Project, Washington, DC: Brookings Institution.

Parliament of Australia, http://www.aph.gov.au/About_Parliament/Parliamentary_Departments/Parliamentary_Library/Browse_by_Topic/ClimateChange/cprs (accessed November 2015).

Parry, Ian W.H. (1995), 'Pollution taxes and revenue recycling', *Journal of Environmental Economics and Management*, **29** (3), S64–S77.

Parry, Ian W.H. (2004), 'Are emissions permits regressive?', *Journal of Environmental Economics and Management*, **47** (2), 364–387.

Pereira, Alfredo M. and Pedro G. Rodrigues (2015), 'A new carbon tax in Portugal: A missed opportunity to achieve the triple dividend?', Working Paper.

Québec Ministère du Développement durable, de l'Environnement et de la Lutte contre les changements climatiques (2015), *Québec's Cap-and-Trade System for Greenhouse Gas Emission Allowances: Technical Overview*, Québec, Canada: Gouvernement du Québec, http://www.mddelcc.gouv.qc.ca/changements/carbone/documents-spede/technical-overview.pdf (accessed November 2015).

RGGI (Regional Greenhouse Gas Initiative) (2015a), *Investment of RGGI Proceeds through 2013*, http://rggi.org/docs/ProceedsReport/Investment-RGGI-Proceeds-Through-2013.pdf (accessed November 2015).

RGGI (Regional Greenhouse Gas Initiative) (2015b), *2015 CO_2 Allowance Allocation*, https://www.rggi.org/docs/CO2AuctionsTrackingOffsets/Allocation/2015_Allowance-Allocation.xls (accessed November 2015).

Snow, Arthur and Ronald S. Warren (1996), 'The marginal welfare cost of public funds: Theory and estimates', *Journal of Public Economics*, **61** (2), 289–305.

Sumner, Jenny, Lori Bird and Hillary Smith (2009), *Carbon Taxes: A Review of Experience and Policy Design Considerations*, National Renewable Energy Laboratory Technical Report NREL/TP-6A2-47312, Golden, CO: NREL.

Triest, Robert K. (1990), 'The relationship between the marginal cost of public funds and marginal excess burden', *American Economic Review*, **80** (3), 557–566.

Weitzman, Martin L. (1974), 'Prices vs. quantities', *Review of Economic Studies*, **41** (4), 477–491.

West, Sarah E. and Roberton C. Williams (2004a), 'Estimates from a consumer demand system: Implications for the incidence of environmental taxes', *Journal of Environmental Economics and Management*, **47** (3), 535–558.

West, Sarah E. and Roberton C. Williams (2004b), *Empirical Estimates for Environmental Policy Making in a Second-Best Setting*, National Bureau of Economic Research Working Paper No. 10330, Cambridge, MA: NBER.

World Bank (2014), *State and Trends of Carbon Pricing*, Report, Washington, DC: World Bank.

14. REDD+: global multilevel forest governance for building a climate resilient society

Abrar Juhar Mohammed and Makoto Inoue

INTRODUCTION

It is now well accepted among academicians and policy makers alike that climate change presents an unparalleled menace to humanity. Efforts to steer policies towards building a climate change resilient society are being given priority. By climate change resilient society, we are referring to a society that contributes to mitigation of climate change, thereby intervening to reduce emissions from the forest and enhancing the ecosystem service as sinks of greenhouse gases (GHG) while having a capacity to adapt to the possible adverse effects of climate change (IPCC, 2014a). Forests play a key role in building a society resilient to climate change by contributing to both climate change mitigation and adaptation. They provide important ecosystem services at the national scale, including carbon sequestration and protection of the life supporting systems that are vital for maintaining human resilience to climate change such as agriculture and hydrology. The world's forests store 289 gigatonnes (Gt) of carbon in their biomass, and 30 million hectares of forest are designated for soil and water conservation, avalanche control, sand dune stabilization, desertification control or coastal protection (FAO, 2010). At the local scale, forest ecosystem services provide safety nets to reduce the vulnerability of forest-dependent communities. For example, non-timber forest products such as fuelwood, water, bush meat, wild fruits and medicines support the livelihood of an estimated 0.25 billion to 1 billion people (Peskett, 2010). Despite their key role in maintaining a healthy biome, forests continue to be lost at an alarming rate. According to FAO (2010), approximately 13 million hectares of forest are lost every year. This deforestation and forest degradation in the tropics contributed to a third of anthropogenic CO_2 emissions between 1750 and 2011 (IPCC, 2014b).

Reducing emissions from deforestation and forest degradation plus

(REDD+) is a global initiative that is being discussed and implemented to conserve and/or enhance the ecosystem services of forests, especially carbon sequestration. REDD+ encompasses 'policy approaches and positive incentives on issues relating to reducing emissions from deforestation and forest degradation in developing countries; and the role of conservation, sustainable management of forests and enhancement of forest carbon stocks in developing countries' (UNFCCC Decision 2/CP.13–11). It is a voluntary market mechanism to ensure that the opportunity cost of reducing carbon emissions by developing nations (Non-Annex I countries) is compensated by developed nations (Annex I countries) (Purnomo et al., 2012). More specifically, it is a mechanism that seeks to create performance-based financial incentives for policies and management actions related to five eligible activities: reducing emissions from deforestation; reducing emissions from forest degradation; conservation of forest carbon stocks; sustainable management of forests; and enhancement of forest carbon stocks (UNFCCC/CP/2010/7/Add.1, para. 70).

The seed of REDD+ was sown in the Kyoto Protocol in 1997 when the role of forests as carbon sinks was recognized (Hiraldo and Tanner, 2011). The Kyoto Protocol mitigation strategy, however, gave major emphasis to the power sector, leaving little space for forestry actions. Nonetheless, governments, conservation non-governmental organizations (NGOs), private consultants, natural scientists and carbon finance companies continued to push for policies to reduce deforestation (Griffiths, 2007). Such advocacy efforts bore fruit in the 11th Conference of the Parties (COP) to the Kyoto Protocol (COP 11/MOP 1) of the United Nations Framework Convention on Climate Change (UNFCCC). In this meeting, the Coalition for Rainforest Nations led by Costa Rica and Papua New Guinea presented a formal proposal for reducing emissions from deforestation (RED). COP 11 requested its Subsidiary Body for Scientific and Technological Advice (SBSTA) to evaluate the issue of reducing emissions from deforestation and report back to the UNFCCC COP 13/MOP 3 in December 2007. At the 2007 COP meeting held in Bali, the parties agreed that tackling forest degradation was as important as tackling deforestation to reduce emissions from forest landscapes. For some countries, forest degradation was noted to be an even bigger problem than deforestation. In addition, it was recognized that, in some cases such as peatland soils, forest degradation can generate higher emission rates. As a result, the parties agreed the need to address both deforestation and degradation together, which led to a modification of RED to 'reducing emissions from deforestation and forest degradation' (REDD) (Griffiths, 2007; Hiraldo and Tanner, 2011).

In the following years, it was further agreed that there could be climate benefits not only from avoiding negative changes (deforestation, degradation) but also from enhancing positive changes, such as conserving and restoring forests. Consequently, these actions were represented by the '+', and 'reducing emissions from deforestation and forest degradation' (REDD) 'and the role of conservation, sustainable management of forests and enhancement of forest carbon stocks in developing countries' (REDD+) became official language at the 2008 COP14 in Poznan (Wertz-Kanounnikoff and Angelsen, 2009).

The recognition and inclusion of co-benefits such as improving community livelihoods, conserving biodiversity and protecting water resources within the REDD+ debate was undertaken in the 2009 COP15 (Hiraldo and Tanner, 2011). A list of safeguards that should be promoted and supported to achieve co-benefits from REDD+ efforts was agreed at the Cancun COP meeting (COP16). The subsequent Durban Agreement also requested the parties implementing REDD+ to provide information on how the safeguards were being addressed and respected throughout the implementation of the REDD+ activities. Further development in the negotiations on the REDD+ framework, termed the Warsaw Framework for REDD+, was achieved at the 19th Conference of the Parties in 2013. The agreed decisions included issues related to finance, national forest monitoring systems, measuring, reporting and verification (MRV) of carbon emissions, and institutional arrangements for REDD+ (UNFCCC, 2013).

At the COP 21 meeting, from the end of November to mid-December 2015, nearly 200 countries signed an agreement to hold the global annual average temperature rise to well below 2°C, with a further effort in pursuing actions to maintain temperature rise at 1.5°C. The text of COP 21 is in two forms: decision and agreements. REDD+ is covered in both texts. In Article 54 of the decision text, under finance, it states the importance of adequate and predictable financial resources, including for results-based payments, for the implementation of REDD+. It also encourages coordination of support from, inter alia, public and private, bilateral and multilateral sources, such as the Green Climate Fund, and alternative sources. In Article 5 of the agreement, it is mentioned that 'Parties should take action to conserve and enhance, as appropriate, sinks and reservoirs of greenhouse gases as referred to in Article 4, paragraph 1(d), of the Convention, including forests.' The second paragraph of the article states that parties are encouraged to take action to implement and support REDD+ (UNFCCC, 2015).

The assumption by REDD+ proponents is that the existing market undervalues the environmental services provided by forests and correcting such market imperfection can enhance society's contribution towards mitigating deforestation and forest degradation (Clements, 2010). Hence, the core approach in REDD+ is to make performance-based payments to forest owners and users to reduce deforestation and forest degradation. Such payments are intended to make forest conservation more lucrative than other alternatives, creating an incentive for forest owners and users to manage forests better and clear less forestland (Angelsen, 2009).

There is significant global interest in REDD+ (Peskett, 2010; Lederer, 2012). This is evident from the global climate-related funding trend and forest carbon market. Out of 14 new international climate-related funds or fund-related initiatives that have been established since 2006, at least seven of them target forests and REDD+ (Griffiths, 2007). The forest carbon market also increased from 30.1 million metric tons of carbon dioxide equivalent ($MMtCO_2e$) in the years 2008 and 2009 to 75 $MMtCO_2e$ in 2010, with an estimated value of $432 million, impacting more than 7.9 million hectares in 49 countries from every region of the world (Diaz et al., 2011). Considering the vital role REDD+ can play in building a climate resilient society and its related influence upon forest and rural society, this chapter provides a comprehensive review of what REDD+ is, how it is implemented and what key issues need to be considered to steer REDD+ towards building a climate resilient society by taking Ethiopia as a case study. Ethiopia is selected because the country represents all three major problems that REDD+ should address towards building a climate resilient society. The country is one of the most vulnerable to climate change (Kreft et al., 2014), while also witnessing a high rate of deforestation. In recognition of the global significance of this ongoing degradation and deforestation, the remnant forests have been incorporated into global biodiversity hotspots by Conservation International (Mohammed and Inoue, 2012).

REDD+ IN ETHIOPIA

The total forest cover of Ethiopia including smallholder/private and state plantations is estimated to be around 13 million hectares, covering some 11.4% of the total land area of the country (FAO, 2010). The major forest types of the country include: moist montane forests; dry evergreen montane forests; *Acacia-Commiphora* woodlands; lowland semi-evergreen tropical forests; combretum-*Terminalia* woodlands; and bamboo thickets (Bekele et al., 2015). These forests have the potential to mitigate the release of 2.76 billion tons of carbon into the atmosphere if Ethiopia protects and sustainably manages its forest resources (Moges et al., cited in Bekele et al., 2015). Nonetheless, the forests continue to diminish at an alarming rate of 150 000–200 000 hectares per year (Zewdu, 2002). In particular, the montane forests, which are among the country's most important homes of biodiversity (Sayer et al., 2005), are located in areas inhabited by the majority of the Ethiopian population. In recognition of the global significance of this ongoing degradation and deforestation, montane forest remnants have been incorporated into global biodiversity hotspots by Conservation International (Mohammed and Inoue, 2012). With population growing at a rate of about 3.19% (2011 estimate) and about 38.7% living below the poverty level (FY05/06 estimate), it is a major challenge to find a system of management that will minimize further destruction of natural forests by balancing protection objectives and utilization interests of the state, non-state organizations and local communities (Zewdu 2002; Mohammed and Inoue, 2012).

Ethiopia is one of the most vulnerable countries in the world to climate change. According to a global-level study by Kreft et al. (2014), Ethiopia was ranked 29th in terms of average number of climate change-related deaths between 1994 and 2013. Amidst the enormous pressure from climate change as well as degradation of the Ethiopian forests, policy measures are being made by the Ethiopian government to harness deforestation and reduce greenhouse gas emissions from deforestation and forest degradation. The country has adopted the Climate-Resilient Green Economy (CRGE) strategy, which aims to build sustainable economic development through substantial CO_2 emission reductions in major economic sectors. The forestry sector is identified as one of the four fast-track implementation pillars responsible for achieving 50% of the national abatement potential. In view of this, the government has fully embraced REDD+ as an integral part of the national CRGE strategy (Bekele et al., 2015).

The first large-scale REDD+ pilot project in Ethiopia is the Bale Mountains Eco-Region REDD+ project. The project builds on the Bale Eco-Region Sustainable Management Programme (BERSMP) that has been running since 2007. The REDD+ project area covers 923 593 hectares, including 576 856 hectares of tropical dry and moist forest. To achieve a reduction in deforestation to 1% in project year 20, participatory forest management (PFM) is being implemented across the Bale Mountains, which includes the establishment of community-based organizations, development of forest management plans, and implementation of sustainable forest management practices. PFM is regarded as a tool for REDD+ implementation alongside additional support for sustainable agricultural intensification, woodlot establishment and improved fire management (Watson et al., 2013). In addition to local people, the main actors in the implementation of this REDD+ pilot project are Oromia Forest and Wildlife Enterprise (OFWE) and NGOs Farm Africa and SOS Sahel Ethiopia. It is expected to run for 20 years. According to the project feasibility study, an estimated 37 MMtCO2e emission reduction is likely to be achieved over the project lifetime (OFWE, 2014).

REDD+ ACTORS AND PROCESS OF IMPLEMENTATION

REDD+ is global multilevel environmental governance regarding how the world's forests are governed (Lederer, 2012). Multilevel governance can be defined as the *processes and structures of public policy, decision-making and management that engage people constructively across the boundaries of public agencies, levels of government, and/or the public, private and civic spheres in order to carry out a public purpose that could not otherwise be accomplished* (Emerson et al., 2012). Multilevel governance is based on the notion of the positive role of interactions and collaborative effort to solve public problems (Inoue et al., 2015). REDD+ is global multilevel governance in that it is the result of vertical and horizontal interaction among actors from the local to the global level. REDD+ allows actors from diverse scales and categories to negotiate, formulate and implement policy (Angelsen, 2009). As argued by Maraseni and Cadman (2015), understanding global multilevel governance in general and REDD+ in particular requires deliberating on structure and process. Structure refers to the players or actors, while the rules of the game and how it is played denote the process.

Government Agencies

In 1992 during the Rio Earth Summit, members of the United Nations adopted three major conventions: the UN Convention on Biological Diversity (UNCBD), the UN Convention to Combat Desertification (UNCCD) and the UN Framework Convention on Climate Change (UNFCCC). Of the three, UNFCCC is the key negotiation platform for REDD+. Representatives from 195 countries including Ethiopia that attended the negotiation, also known as Parties to the Convention, signed the agreements. The convention divides these countries into three: Annex I, Annex II and Non-Annex I. Annex I countries are industrialized countries and countries with economies in transition, including the Russian Federation, the Baltic states, and several Central and Eastern European states, while the industrialized countries alone form Annex II Parties. The Non-Annex I Parties are mostly developing countries, to which Ethiopia belongs. The highest decision-making body of UNFCCC, the Conference of the Parties (COP), is formed by representatives of Parties to the Convention. REDD+ and the agreement signed by Parties to the Convention, which is a base for REDD+ implementation, are a product of negotiation through a series of COP meetings (UNFCCC, 2015).

In addition to their role as negotiators at the global level, governments play an important role in the implementation of REDD+. Many developing countries are directing their forest policy in line with the objectives of REDD+ (Mohammed et al., 2016). Countries from the Annex I category are mainly involved in providing financial resources for the implementation of forest conservation policies in Non-Annex I countries. For example, Norway and the UK provided about 10 million USD for supporting REDD+ readiness in Ethiopia. Subnational-level actors have also been instrumental in the implementation of REDD+. In fact, although REDD+ was conceived as a national approach, many of the early advances are being realized at the subnational level (Ravikumar et al., 2015). In Ethiopia, the main actor in the implementation of the first and largest REDD+ project, the Bale Mountains Eco-Region REDD+ project, is a subnational actor named Oromia Forest and Wildlife Enterprise (Figure 14.1).

Funding (Donor)Agencies

The seven major funding sources for REDD+ are the World Bank Forest Carbon Partnership Facility (FCPF), the World Bank Forest Investment Programme (FIP), the UN REDD Programme, the GEF Tropical Forest Account, the Congo Basin Forest Fund, the Norwegian Forest and Climate Initiative, and the International Forest Carbon Initiative (Asia-Pacific). In most cases, these funds aim to support national-level REDD readiness of developing countries' governments and/or project-level pilot and demonstration REDD+ activities (Mohammed and Inoue, 2016). Among these funding agents, the FCPF is playing an important role in the implementation of REDD+ in Ethiopia.

The FCPF is administered by the World Bank, a multilateral financial institution. The governing boards of the FCPF include representatives of REDD+ recipient countries and donors. The FCPF Participants Committee grants observer status to non-donor stakeholders, including the private sector, NGOs and forest-dependent indigenous peoples and other forest dwellers. The primary scale and scope of focus of the FCPF are national-level REDD+ readiness, including carbon mitigation and safeguarding. The FCPF is currently working on such activities with 37 partner countries (McDermott et al., 2012). Ethiopia is one of these partner countries and has received 200000 USD from the FCPF for facilitating REDD+ readiness (Figure 14.1).

Certifying Agents

The third set of key actors are those involved in certifying REDD+ projects. The two certification standards that have undertaken the greatest verification of REDD+ initiatives are Verified Carbon Standard (VCS) and Climate Community and Biodiversity Alliance (CCBA). The standards developed by CCBA, VCS or other certifying agents are used by an independent, approved auditor to determine conformance of the project with the standards. These independent bodies might be intentional non-profit organizations (e.g. Rainforest Alliance) or profit organization (e.g. TÜV SÜD South Asia Pvt Ltd). The mission of VCS, among others, is to deliver quality assurance to voluntary carbon markets and pioneering innovative rules and tools for the carbon market.[1] Unlike VCS, CCBA does not engage in carbon accounting for the purpose of selling verified carbon credits. In both CCBA and VCS, a project gets verification after independent, approved auditors determine conformance with the standards in two stages, validation and verification (CCBA, 2013). Both the VCS and CCBA standards are used to verify the Bale Mountains Eco-Region REDD+ project of Ethiopia.

Experts and the Scientific Community

The most influential role of experts and the scientific community in the design and implementation of REDD+ is through their influence on COP decisions. The UNFCCC has two permanent subsidiary bodies: the SBSTA and the Subsidiary Body for Implementation (SBI). They are both open to participation by any party. Governments often send representatives who are experts in the fields of the respective bodies. The SBSTA provides the COP with advice on scientific matters such as guidelines for preparing national emission inventories and methodological works in specific areas including adaptation and vulnerability. The SBI gives advice to the COP on all matters concerning the implementation of the Convention.

The Intergovernmental Panel on Climate Change (IPCC) is another channel through which experts and the scientific community influence the making of REDD+. There are several research centres, institutes and universities that are engaged in developing knowledge about climate change, climate change mitigation and climate change adaptations. The IPCC is a scientific body composed of thousands of these scientists from all over the world contributing on a voluntary basis as authors, contributors and reviewers. The scientists review and assess the most recent scientific, technical and socio-economic information produced worldwide relevant to the understanding of climate change.[2] The IPCC works closely with the SBSTA.

In addition to their influence at the global level, the scientific community play a significant role in national- and local-level REDD+ implementation through policy research and evaluation. They also serve as independent consultants during the preparation of REDD+ readiness proposals or implementation of REDD+ pilot projects. The Nature Conservation Research Centre (NCRC) of Ghana, the Institute for Conservation and Sustainable Development of Amazonas (IDESAM) of Brazil, and Oxford University Environmental Change Institute, Centre for Tropical Forests of the UK, for example, were providers of technical consultancy for the preparation of the Bale Mountains Eco-Region REDD+ project of Ethiopia.

NGOs

Non-governmental organizations' role in REDD+ can be traced back to the very genesis of REDD+. It was the tireless advocacy effort by NGOs, together with some national governments such as the Coalition for Rainforest Nations, that brought the REDD+ negotiation to the UNFCCC table. Since then they have been active in the negotiations and implementation of the policy. Over 1598 NGOs are admitted to attend sessions of the COP and its subsidiary bodies as observers. These NGOs represent, among others, indigenous populations, women's and gender groups, and environmental groups (UNFCCC, 2015).

At the national and local level, NGOs are involved as partners to develop, implement, monitor and evaluate REDD+ activities. They also participate in addressing the capacity constraints of all countries involved, especially those of Non-Annex I countries. In Ethiopia, almost all REDD+ projects are being developed by NGOs. Farm Africa and SOS Sahel are the two NGOs that are playing a major role in the development and realization of the Bale Mountains Eco-Region REDD+ project. In fact, as early as 2006, REDD+ was solely the concern of the two prominent NGOs before it was embraced by the Ethiopian government. In addition, the two NGOs were involved in providing advisory services on the development of REDD+ during the R-PP development by the EPA. Other important roles of NGOs in the implementation of REDD+ are the development of standards (such as those discussed in the section 'Certifying agents') and acting as auditors for certification.

Private Sector

The private sector stakeholders play a greater role in REDD+ implementation, mainly in the carbon marketplace, as investors and project developers. According to data from ecosystem services cited in Diaz et al. (2011), 19.1 $MMtCO_2e$ out of the total 29.0 $MMtCO_2e$ were contracted to private buyers. Major private firms like BNP Paribas, Nedbank Group, Gazprom Marketing and Trading (GM&T), the Macquarie Group and others are also involved in both the supply side, by providing direct investment, and the demand side, by purchasing agreement for profit credit or for fulfilling carbon commitment. At the local level too, private land owners seem to have gained momentum in recent years as the preferred site for REDD+ project development. According to the study by Diaz et al. (2011), 62% and 77% of projects by non-profit and for-profit project developers respectively were developed in areas including private landholding, while the percentage of communal land was only 30% and 25% respectively. Finally the private sector is also involved in certification of REDD+ projects. Among others, this includes Asociación Española de Normalización y Certificació (AENOR), Environmental Services, Inc, RINA Services and TÜV SÜD South Asia Pvt Ltd. The latter is an auditor for the Bale Mountains Eco-Region REDD+ project.

Local Communities and Indigenous People

Forests are home to about 300 million people around the world, of whom a third are indigenous peoples. These people are almost entirely dependent on forests for their survival. In addition, about 800 million people living in and around forests rely on them for fuel, food or subsistence income (FCPF, n.d.). Core safeguards devised to secure positive social impacts of REDD+ among indigenous peoples, including the equitable distribution of benefits and free prior and informed consent (FPIC), were agreed at the Cancun COP meeting (COP 16). Main REDD+ project readiness funders such as FCPF and UN-REDD also recognized the important role of local communities and indigenous people throughout REDD+ project design and implementation. Both have expressed their commitment to support indigenous engagement in national and international initiatives at all levels of the process. As a result, all 26 REDD+ readiness proposals presented to the FCPF make reference to community forest management. In some countries such as Ethiopia, Madagascar, Nepal, and the United Republic of Tanzania a nationally organized community management programme is central to the proposed national REDD+ approach (Skutsch and McCall, 2012). In the Bale Mountains Eco-Region REDD+

project, communities living in and around the forest are the main actors playing a vital role in realizing the project's goal. As of 2013, there were 58 forest cooperatives engaged in participatory forest management. The cooperatives manage forest concession areas through legally binding forest management agreements with the government. These stipulate the benefit sharing arrangements between the government (OFWE) and the participating communities. The future plan is to establish a Forest Union that can organize these different cooperatives under one coordinating body that can also engage in advocacy and lobbying on behalf of its members.

IMPLEMENTATION PROCESS

An increasingly accepted approach for the implementation of REDD+ is through three possibly overlapping phases: the 'readiness' phase, 'more advanced readiness' and 'full-scale implementation' (Meridian Institute, 2009). The implementation of REDD+ in Ethiopia has also followed the same phases (Figure 14.1). Funding for the first and second phases is not necessary to be associated with funding for carbon markets, while Phase 3 is linked to payments for quantifiable tons of CO_2 emission reductions (Streck, 2010).

The first 'readiness' phase of Ethiopia can be tracked back to the establishment of the Climate Change Forum–Ethiopia (CCF-E) through a collaborative effort of policy makers and stakeholders. The CCF-E was responsible for coordinating actions, communicating information, convening policy dialogue and initiating public campaigns about climate change in the country. It also led to the development of a draft national climate change strategy. These activities are supported by voluntary contributions such as the World Bank's Forest Carbon Partnership Facility, UN-REDD and other bilateral arrangements. Eligibility for access to funds is based on fulfilment of the requirement by the funding organization. For example, in the FCPF phased procedure, a country needs first to submit REDD concept notes (R-PINs) that provide an overview of land use, deforestation causes, stakeholder consultation and institutional issues for addressing REDD and to summarize countries' requests for assistance to prepare for their REDD programmes. The R-PIN preparation is not funded. However, the FCPF supports up to \$200 000 for the next phase, the readiness preparation plan (R-PP). Ethiopia, through the REDD+ Secretariat housed in the Environmental Protection Authority (EPA), first submitted the R-PIN in 2008, followed by a draft R-PP in 2010. The final R-PP was submitted in 2011.The preparation of the R-PP was supported by a \$200 000 grant from FCPF. The R-PP document sets out the steps and requirements for

Readiness

Government agencies Government of Ethiopia	***Role*** Prepare R-PIN and R-PP and submit to FCPF in 2008 and 2012 respectively
Donor agencies FCPF Norway and DFID (UK)	USD 200000 USD 10 million

Advanced readiness

Government agencies National REDD + Secretariat (NRS)	***Role*** R-PP implementation (2013–2016) Legal and institutional study MRV system establishment Baseline on trends and rates of deforestation and forest degradation Drivers study Guidelines for grievance and conflict resolution Carbon benefit sharing framework
Subnational actors Oromia Forest and Wildlife Enterprise (OFWE)	Implementation of pilot REDD + project (Bale Mountains Eco-Region REDD + project)
NGOs Farm Africa and SOS Sahel	Implementation of pilot REDD + project (Bale Mountains Eco-Region REDD + project)
Donor agencies Government of Norway	EUR 2 million for implementation of pilot REDD + project (Bale Mountains Eco-Region REDD + project)
Certifying agents VCS and CCBA	Developed the VCS and CCBA standards to be used by the auditor for checking conformity of the project with certain standards
Private sector TÜV SÜD South Asia Pvt Ltd	An auditor of the Bale Mountains Eco-Region REDD + project using the VCS + CCBA standards
Scientific community As a consultant NCRC, IDESAM, Oxford University – Environmental Change Institute, Centre for Tropical Forestry	Consultation on project development, baseline deforestation modelling
Local communities People living in and around the project area	Benefit sharing negotiation, taking forest protection and management responsibilities

Compliance

Private Sector Probably large-scale companies	***Role*** Buy carbon credits to compensate for their carbon emissions
Governments Developed countries	Norway has pledged to purchase verified emission reduction
Governments Government of Ethiopia	Negotiate for selling carbon credits, benefit sharing within the country, implement REDD + post-compliance phase
Local communities	Contribute for protection and management of forest, obtain (fair?) financial benefit

Figure 14.1 Actors and implementation phases of REDD+ in Ethiopia

Ethiopia to achieve 'readiness'. This included consultation and an outreach plan and rapid analysis of land use, forest policy and governance assessment, a REDD implementation framework, and terms of reference for social and environmental impact assessment of the REDD strategy (Griffiths, 2007). Pilot projects such as the Bale Mountains Eco-Region REDD+ project were also identified at the R-PP preparation stage.

The second phase is 'more advanced readiness', where implementation of policies and measures (PAMs) to reduce emissions is undertaken. The implementations include land tenure and governance reforms, forest law enforcement, forest management planning, reduced impact logging, expansion of forest reserves, wildfire and pest management, and others. This phase can be funded in two stages, the first for implementation of measures and the second for measuring carbon storage (Griffiths, 2007). After the endorsement of the R-PP, an R-PP implementation grant of $3.6 million (including R-PP preparation grant of $200 000) was approved by FCPF. In 2013, Ethiopia entered the advanced readiness phase. Since then, the REDD+ Secretariat has been responsible for the overall implementation of the readiness process outlined in the R-PP. Additional funding of $10 million was provided by the governments of Norway and the UK for the implementation of the R-PP. At the national level, planning for developing a system for measuring, monitoring, reporting and verifying changes in forest cover and forest carbon stocks has started, and an MRV roadmap has been developed. In addition, the Food and Agriculture Organization (FAO) of the United Nations, together with the then Ministry of Agriculture (MoA), now the Ministry of Environment and Forests (MoEF), has started work on developing a national forest inventory.

The third phase is full UNFCCC 'compliance'. This stage involves a results-based compensation mechanism for fully measured, reported and verified emission reductions and removals from the forestry sector (Meridian Institute, 2009). Data on forest emissions reduced or carbon stocks enhanced from each forest are collected through a national, regional or international MRV system and submitted to a national REDD+ payment authority, a UNFCCC institution and/or international buyers of REDD+ credits to earn payments (Wertz-Kanounnikoff and Angelsen, 2009).

OUTCOMES SO FAR

REDD+ has transformed the institutional, socio-economic and ecological outcomes from the forestry sector since its commencement in 2005

(Mohammed and Inoue, 2016). Here we discuss them with special emphasis on Ethiopia.

Institutional

REDD+ has been instrumental in shaping the forestry institution through its 'readiness' procedure, which demands that countries, mainly Non-Annex I, demonstrate strong institutional capacity (Mohammed and Inoue, 2016). In Ethiopia, the issue of climate change in general and REDD+ in particular had a profound impact on the attention given to forests and the environment. Despite a long history of forest professional advocacy for a dedicated federal-level forestry institution, it remained under the aegis of agriculture until 2013, when the MoEF was established. The timing of the establishment of the MoEF with that of the commencement of implementation of the R-PP is no coincidence. It was as part of the institutional strengthening of the forestry sector. REDD+ influenced not only the structure but also the process of agenda setting and policy formulation. Unlike the previous policy process, where consultation was superficial, the REDD+ process has been mainly based on consultation of wider stakeholders during the development and implementation of the R-PP. For example, a total of 24 consultation workshops were conducted across the country, including two national-level workshops, seven regional-level workshops, one zonal workshop, seven workshops at the district level and seven community consultations with forest-dependent peoples in seven regional states. However, the forest-dependent communities, key stakeholders who need to be engaged in REDD+ implementation, have so far been less involved (Bekele et al., 2015).

The other important institutional issues in the forestry sector of developing nations in general and for the implementation of REDD+ in particular is tenure clarity. In most of the regions where REDD+ is being implemented, tenure right to land remains unclear. Indigenous and local people customarily own about 65% of the total land. However, only 18% of this land is legally recognized (RRI, 2015). Consequently, there are both hopes and fears about what the REDD+ mechanism can achieve (Bolin and Tassa, 2012). At the project level, the likes of the Bale Mountains Eco-Region REDD+ project that are built on previously implemented community-based projects tend to have the upper hand, as the tenure issue is clarified through agreements signed between government and the communities. However, for the majority of the forests of developing countries in general, and Ethiopia in particular, the tenure issue remains blurred. The implementation of the R-PP is hoped at least to minimize this problem.

Currently, nearly all natural forests and dry woodlands are under state control as state property, alienating communities from the forests they have been using for centuries. There is hope, though, as the proposed forest law under revision hints at the possibility of recognizing community forests as a separate property-rights regime.

Socio-economic

Since the commencement of REDD+, the value of standing trees as a carbon sink has increased significantly. The market to buy this carbon has also been increasing (Diaz et al., 2011). However, whether REDD+ can make conserving forests financially competitive remains contentious. At the local level, in order to be competitive, REDD+ needs to fully recognize and compensate for forgone livelihood opportunities of forest-based communities affected by the policy implementation. Such a calculation is quite difficult, because the livelihood activity forgone may be a source of survival for many other people living far from the forest. Moreover, the wider economic benefits of forest land uses that accrue through support of processing, marketing and service industries are usually not considered in the REDD+ opportunity forgone calculation (Ghazoul et al., 2010). In Ethiopia, so far there is no cost and benefit sharing mechanism in place for the national REDD+ programme. The R-PP document notes the need to improve local people's livelihoods though income diversification. Moreover, at the project level, the possibility of a 60/40 share in which the community will get more than half of the REDD+ income is being discussed. Whether that will be agreed and implemented, as well as whether it will be enough to compensate for cost forgone by depriving local people of forest goods, will be key for sustainable REDD+ implementation.

Ecological

As the main objective of REDD+ is avoiding deforestation and forest degradation, it is expected to have a positive ecological outcome. There is, however, an important concern with respect to the implications of REDD+ implementation for biodiversity conservation. The choice of forest areas to be protected for emission reduction benefits might not always coincide with areas of high biodiversity value (Miles and Kapos, 2008). Scholars such as Huettner (2012) have expressed concerns about potential land use pressure on high biodiversity but low carbon areas, including conversion of biodiversity-rich forest and non-forest land into plantations. Others, including Phelps et al. (2012), are wary about potential diversion of funds for forest conservation away from high biodiversity, low carbon areas. In Ethiopia, project-level REDD+ implementation, so far, has targeted forests with both high biodiversity and carbon sequestration. The Bale Mountains Eco-Region, where the REDD+ pilot project is being implemented, has high biodiversity and endemism, with a quarter of mammals and 6% of birds in the area being Ethiopian endemics. There are also a number of rare endemic amphibians. The area contains the entire global population of the giant molerat (*Tachyoryctes macrocephalus*), and the largest global populations of endangered Ethiopian wolves (*Canis simensis*) and mountain nyala (*Tragelaphus buxtoni*). The Bale Mountains also harbour 40% of the 1000 known species of medicinal plants in Ethiopia, and the largest remaining natural stand of wild coffee genetic stock. The park includes an Afroalpine plateau over 3500m asl, which is the largest area of Afroalpine habitat on the African continent, as well as a major section of moist tropical forest, the second largest in Ethiopia. Owing to this, the Bale Eco-Region is one of the eight World Heritage Sites of Ethiopia registered by UNESCO. Hence the effort to conserve the forest habitat through the REDD+ scheme will also contribute to the conservation of a large biodiversity pool. Further, it will protect some of the most endangered species in the world.

CONCLUSIONS

Building a climate change resilient society requires steering policies to enhance society's efforts to mitigate climate change while developing capacity to adapt to the adverse effects of climate change. One major recent global initiative targeted to contribute to both mitigation and adaptation, especially for developing countries, is REDD+. Since submitting the Readiness Plan Idea Notes (R-PINs) in 2008, Ethiopia has been actively involved in implementing REDD+ policy. A total of seven major actors, that is, government agencies, funding (donor) agencies, certifying agents, experts and the scientific community, NGOs, the private sector, and local and indigenous people, are identified to play an important role in the implementation of REDD+ in Ethiopia. Ethiopia is following the three-phase implementation process of 'readiness' phase, 'more advanced readiness' and 'full-scale implementation'. The country entered the more advanced readiness stage, which involves implementation of the readiness preparation plan (R-PP) that is prepared during the readiness phase, in 2013. Among other things, the country is implementing a REDD+ pilot project, including the pioneer Bale Eco-Region REDD+ project.

The most important impact of REDD+ so far has been on the economic and institutional front. In Ethiopia, the long-sought federal-level forestry organ was inaugurated in 2013, thanks to the requirements of REDD+ funding agencies. Moreover, the process of REDD+ policy document preparation has been more participatory than the previous forestry policy formation. On the economic front, the government was able to get millions of dollars through bilateral arrangements as well as from multilateral financial institutions such as FCPF. Since the full-scale implementation stage where carbon is sold is not yet achieved, whether the income from REDD+ will be equitably shared with local people is yet to be seen. So far there are different schemes, including a 60/40 share basis, whereby local people will get 60% of the total sale being considered. In order to get local people's participation for sustainable implementation of REDD+, as well as enhance their capacity to cope with the potential impact of climate change, we argue that the lion's share of the income from carbon sale should be channelled to the local people. Although many scholars have expressed their concerns about the potential adverse biodiversity outcome of REDD+, currently its implementation in Ethiopia seems to favour not only carbon sequestration but also biodiversity conservation.

NOTES

1. See http://www.v-c-s.org/who-we-are/mission-history.
2. See http://www.ipcc.ch/organization/organization_structure.shtml.

REFERENCES

Angelsen, A. (2009), 'Introduction', in A. Angelsen (ed.), *Realising REDD+: National Strategy and Policy Options*, Bogor, Indonesia: CIFOR, pp. 1–9.

Bekele, M., Y. Tesfaye, Z. Mohammed, S. Zewdie, Y. Tebikew, M. Brockhaus and H. Kassa (2015), *The Context of REDD+ in Ethiopia: Drivers, Agents and Institutions*, Occasional Paper 127, Bogor, Indonesia: CIFOR.

Bolin, A. and D.T. Tassa (2012),'Exploring climate justice for forest communities engaging in REDD+: Experiences from Tanzania', *Forum Development Studies*, **39**, 5–29.

CCBA (2013), *Climate, Community and Biodiversity Standards*, 3rd edn, Arlington, VA: CCBA, www.climate-standards.org.

Clements, T. (2010), 'Reduced expectations: The political and institutional challenges of REDD+', *Oryx*, **44**, 309–310.

Diaz, D., K. Hamilton and E. Johnson (2011), *State of the Forest Carbon Markets 2011: From Canopy to Currency*, Washington, DC: Ecosystem Marketplace, Forest Trends.

Emerson, K., T. Nabatchiand and S. Balogh (2012), 'An integrative framework for collaborative governance', *Journal of Public Administration Research and Theory*, **22** (1),1–29.

FAO (Food and Agriculture Organization) (2010), *Global Forest Resources Assessment 2010*, Rome: FAO.

FCPF (Forest Carbon Partnership Facility) (n.d.), *Demonstrating Activities That Reduce Emissions from Deforestation and Forest Degradation*, Washington, DC: World Bank.

Ghazoul, J., R.A. Butler, J. Mateo-Vega and L.P. Koh (2010), 'REDD: A reckoning of environment and development implications', *Trends in Ecology and Evolution*, **25**, 396–402.

Griffiths, T. (2007), *Seeing RED? Avoided Deforestation and the Rights of Indigenous Peoples and Local Communities*, Moreton-in-Marsh, UK: Forest Peoples Programme.

Hiraldo, R. and T. Tanner (2011), 'Forest voices: Competing narratives over REDD+', *IDS Bulletin*, **42**(3), 42–51.

Huettner, M. (2012) 'Risks and opportunities of REDD+ implementation for environmental integrity and socio-economic compatibility', *Environmental Science and Policy*, **15**, 4–12.

Inoue, M., G.P. Shivakoti and H.R. Ojha (2015), 'Multi-level forest governance in Asia: An introduction', in M. Inoue and G.P. Shivakoti (eds), *Multi-level Forest Governance in Asia: Concepts, Challenges and the Way Forward*, New Delhi: Sage, pp. 1–18.

IPCC (2014a), *Climate Change 2014: Working Group III: Impacts, Adaptation, and Vulnerability*, Cambridge, UK: Cambridge University Press.

IPCC (2014b), *Climate Change 2014: Working Group III: Mitigation of Climate Change*, Cambridge, UK: Cambridge University Press.

Kreft, S., D. Eckstein, L. Junghans, C. Kerestanand and U. Hagen (2014), *Global Climate Risk Index 2015: Who Suffers Most From Extreme Weather Events? Weather-Related Loss Events in 2013 and 1994 to 2013*, Bonn: German watch.

Lederer, M. (2012), 'REDD+ governance', *WIREs Climate Change*, **2012** (3), 107–113.

Maraseni, T.N. and T. Cadman (2015), 'A comparative analysis of global stakeholders' perceptions of the governance quality of the clean development mechanism (CDM) and reducing emissions from deforestation and forest degradation (REDD+)', *International Journal of Environmental Studies*, **72** (2), 288–304.

McDermott, M., S. Mahanty and K. Schreckenberg (2012), 'Examining equity: A multidimensional framework for assessing equity in payments for ecosystem services', *Environmental Science and Policy*, **33**, 416–427.

Meridian Institute (2009), *REDD+ Institutional Options Assessment: Developing an Efficient, Effective, and Equitable Institutional Framework for REDD+ under the UNFCCC*, http://www.REDD-OAR.org.

Miles, L. and V. Kapos (2008), 'Reducing greenhouse gas emissions from deforestation and forest degradation: Global land-use implications', *Science*, **320**, 1454–1455.

Mohammed, A.J. and M. Inoue (2012), 'Drawbacks of decentralized natural resource management: Experience from Chilimo Participatory Forest Management project, Ethiopia', *Journal of Forest Research*, **17** (1), 30–36.

Mohammed, A.J. and M. Inoue (2016), 'Understanding REDD+ with actor-centered power approach: A review', *Journal of Biodiversity Management and Forestry*, **5** (1), 1–8.

Mohammed, A.J., M. Inoue, R.J. Peras, T.K. Nath, M. Jashimuddin and J.M. Pulhin (2016), 'Transformation strategy for managing coupled socio-ecological

systems: Case studies from Bangladesh and the Philippines', *Small Scale Forestry*,**15** (2), 213–227.
OFWE (2014), 'Bale Mountains Eco-Region Reduction of Emission from Deforestation and Forest Degradation (REDD+) Project', Project report submitted to CCBA.
Peskett, L.with J. Stephenson (2010), *Is REDD+ an Opportunity to Support Climate Compatible Development in Developing Countries?*, CDKN Policy Brief, London: Climate and Development Knowledge Network.
Phelps, J., E.L. Webb and W.M. Adams (2012), 'Biodiversity co-benefits of policies to reduce forest-carbon emissions', *Nature Climate Change*, **2**, 497–503.
Purnomo, H., G. Arum, R. Achdiawan and R. Irawati (2012), 'Rights and wellbeing: An analytical approach to global case comparison of community forestry', *Journal of Sustainable Development*, **5** (6), 35–45.
Ravikumar, A., A.M. Larson, A.E. Duchelle, R. Myers and J.G. Tovar (2015), 'Multilevel governance challenges in transitioning towards a national approach for REDD+: Evidence from 23 subnational REDD+ initiatives', *International Journal of the Commons*, **9** (2),909–931.
RRI (Rights and Resources Initiative) (2015), *Who Owns the World's Land? A Global Baseline of Formally Recognized Indigenous and Community Land Rights*, Washington, DC: RRI.
Sayer, J., C. Elliott, E. Barrow, S. Gretzinger, S. Maginnis, T. McShane and G. Shepherd (2005), 'Implication for biodiversity conservation of decentralized forest resource management', in C.J. Pierce Colfer and D. Capistrano (eds), *The Politics of Decentralization: Forests, People and Power*, London: Earthscan, pp. 121–138.
Skutsch, M. and M.K. McCall (2012), 'The role of community forest management in REDD+', *Unasylva*, **239** (63), 51–56.
Streck, C. (2010), 'Reducing emissions from deforestation and forest degradation: National implementation of REDD schemes', *Climatic Change*, **100**, 389–394.
UNFCCC (2013), Decision from the Nineteenth Session of the Conference of the Parties (COP 19), Warsaw, 11–22 November, http://unfccc.int/meetings/warsaw_nov_2013/session/7767/php/view/decisions.php.
UNFCCC (2015), 'Parties and observers', http://unfccc.int/parties_and_observers/items/2704.php (accessed October 2015).
Watson, C., S. Mourato and E.J. Milner-Gulland (2013), 'Uncertain emission reductions from forest conservation: REDD in the Bale Mountains, Ethiopia', *Ecology and Society*, **18**, 3–6.
Wertz-Kanounnikoff, S. and A. Angelsen (2009), 'Global and national REDD+ architecture: Linking institutions and actions', in A. Angelsen (ed.), *Realising REDD+: National Strategy and Policy Options*, Bogor, Indonesia: CIFOR, pp. 13–24.
Zewdu, M. (2002), *Sustainable Development in Ethiopia: Report of Assessment of Activities and Issues Relevant to the Review Process of the Earth Summit 2002 in Ethiopia*, Addis Ababa: Institute of Biodiversity Conservation and Research.

15. Democratising climate finance at local levels

Victor Orindi, Yazan Elhadi and Ced Hesse

INTRODUCTION

Although the total annual flow of climate finance is difficult to establish, as there is no definitive definition of what climate finance includes (Gupta et al., 2014),[1] global climate finance reaching developing countries is set to increase with governments from developed countries setting the collective target of US$100 billion annually to be channelled to developing countries by 2020 through the Green Climate Fund (GCF) (UNFCCC, 2015). The commitment of the GCF to direct 50% of its anticipated US$100 billion per year for climate change adaptation offers significant opportunities for investment in climate resilient development pathways.

To be effective in supporting climate adaptation, climate finance must reach and be prioritised by the communities that need it the most and be used to fund solutions that work on the ground (Steele et al., 2015). To achieve this, mechanisms are needed that channel climate finance from the global and national levels to local communities in a way that is transparent, participatory and efficient. This is proving to be a major challenge. Evidence points to the concentration of climate finance at the national level with insufficient recognition of climate finance needs at sub-national and local levels. Finance is often channelled to developing countries based on the assumption that funds disbursed to national accounts will 'trickle down' to all levels. In reality, this goal is not always being met (Christensen et al., 2012).

The GCF offers an opportunity to address this structural blockage. Its 'direct access' modality offers public and private domestic institutions that have been accredited as national implementing entities (NIEs) the opportunity to directly draw down finance from the GCF and to have full control over the programming of how and on what the funds are spent. Such control could extend to establishing mechanisms to enable sub-national bodies to receive and prioritise climate finance in support of local priorities in adaptation and mitigation.

In Kenya, the introduction of devolved government in 2013 provides a potential 'ready-made' institutional architecture for delivering climate finance from the national level to the local level. County governments, with the powers conferred to them by the Constitution of Kenya of 2010, could ensure the delivery of climate finance from the national level to county, sub-county and ward level – the lowest governance/planning unit.[2] Using existing devolved government structures offers good value for money, while also contributing to the realisation of the concepts and principles of democratic governance as enshrined in the Constitution (Mapesa and Kibua, 2008).

The National Drought Management Authority (NDMA) in Kenya, in anticipation of becoming an accredited NIE of the GCF, is piloting the establishment of County Climate Change Funds (CCCFs) to prepare county governments to access climate finance and mainstream climate change into planning for adaptation and climate resilient development. Initially piloted in Isiolo County in 2012, under the then Ministry of State for Development of Northern Kenya and Other Arid Lands, the NDMA, with the technical support of the Adaptation Consortium,[3] has expanded the pilot to the arid and semi-arid counties of Kitui, Makueni, Wajir and Garissa, which together represent approximately 29% of Kenya's land area.

The purpose of this chapter is to present the experiences of the NDMA in piloting the CCCF approach in Isiolo County in Kenya. In the next section, the CCCF's objectives and approach are described, while the section after that presents key achievements. The chapter then presents lessons learned and conclusions.

AN INTEGRATED APPROACH TO DEMOCRATISED CLIMATE FINANCE

The approach piloted by the NDMA with the support of the Adaptation Consortium is designed to establish mechanisms not just to give county government access to climate finance and to integrate climate information into county planning and budgeting systems, but also to allow poor and vulnerable households to prioritise investments that will provide resilient pathways out of poverty and climate vulnerability. The approach, summarised in Figure 15.1 (p. 256), will enable county governments to institutionalise a decision-making process that puts communities in control of their adaptation priorities, and it will help to ensure that vulnerable communities can oversee the flow of climate finance from national to local level, reducing the risk of elite capture for political or personal gain. The work is being implemented within the framework of devolved governance provided by the Constitution of Kenya that obliges county governments to ensure citizen-led and rights-based approaches to the planning and prioritisation of public funding for development.

The approach consists of four interrelated elements that are core to the model and function as explained below.

County Climate Change Fund

The CCCF is a public fund under the discretionary management of the county government, with the necessary fiduciary mechanisms to ensure accountability and transparency. The fiduciary mechanisms are consistent with public finance policy and law and complement the counties' existing finance systems. As public funds, the CCCFs can be capitalised from various sources, such as county development budgets, national climate funds or in-country bilateral and multilateral donors. It is anticipated that the counties will be able to draw down resources from these global climate funds. The CCCFs are designed primarily to finance local adaptation. In this respect, 70% of the CCCF is earmarked to finance ward-level investments, while 20% is allocated to county-level investments and 10% assigned for the administrative running costs of the fund and the adaptation planning committees at ward and county level. The ward is the lowest planning unit under the county government. Fiduciary standards and safeguards are described in the CCCF procedures manual. The manual was developed in consultation with the adaptation committees at county and ward level (see below) and finance and project management specialists, and was then certified by an auditor as meeting donor standards while being consistent with the public finance management system. The manual details requirements for reporting, monitoring and evaluation, and accountability.

Adaptation Planning Committees at Ward and County Levels

Representative ward adaptation planning committees (WAPCs) are responsible for the identification and prioritisation of investments in local public goods that strengthen the adaptive capacities of communities. Using the administrative funds at their disposal (i.e. their share of the 10% of the CCCF allocated for administration), WAPCs conduct participatory assessments of a community's vulnerability or resilience to climate hazards and future climate change. These assessments are then used by the WAPCs through a community consultation process to prioritise investments in public goods whose costs fall within their budget envelope and which meet

> **BOX 15.1: CCCF FUNDING CRITERIA**
>
> 1. Must benefit many people.
> 2. Must support the economy, livelihoods or important services on which many people depend.
> 3. Must be relevant to building resilience to climate change.
> 4. Must encourage harmony, and build relations, understanding and trust.
> 5. Must have been developed after consultation with all potential stakeholders.
> 6. Must be viable, achievable and sustainable.
> 7. Must be cost effective and give value for money.

the funding criteria that promote climate resilient growth and adaptive livelihoods (see Box 15.1).

Each ward committee is allocated a set budget that represents an equal share of the overall budget allocation earmarked for ward-level investments (i.e. 70% of the CCCF). Providing ward committees with a known budget against which to prioritise investments encourages a more effective and participatory planning process. Committee members consider options and weigh up the costs and benefits of different investments against the CCCF funding criteria and arrive at a consensual decision. It encourages active participation by committee members, as they know that their chosen priorities will be funded as long as they meet the fund's criteria and fall within their budget allocation. This is a major factor of motivation. And it supports more careful planning, as committee members know they will be held accountable for their choices. Funds are divided equally among wards, rather than according to population density or vulnerability. This is in recognition that in arid and semi-arid environments, characterised by high climate variability, communities are mobile and typically access local public goods such as water or forest resources across different ecological and administrative boundaries.

The prioritised investments are submitted for review to the county adaptation planning committee (CAPC) composed of representatives from the ward committees, local government and other stakeholders. The CAPC does not have authority to reject WAPC prioritised proposals if the first five proposal criteria are met (see Box 15.1). They are expected to provide additional technical support to the WAPCs, and work together to ensure the proposals meet the last two criteria. Once approved, WAPCs then negotiate and sign contracts with service providers based on phased payments from the CCCF and guided by the Public Finance Management Act 2012.

This process of decision-making enables local people, through their ward committees, to remain in control of their development and adaptation priorities in keeping with the provisions of the Constitution and the County Governments Act. Critically, higher levels of government cannot veto, but only work to strengthen, ward-level proposals. Furthermore, ward committees by managing the tendering process, which is often open to political and economic abuse, are better able to ensure, and account for, the good use of their allocation of the CCCF. This builds local accountability and transparency.

County-level adaptation committees are also responsible for prioritising investments in public goods that benefit the county as a whole. These investments are funded from the 20% allocation of the CCCF and might include projects that support the dissemination of climate information services, or the provision of marketing or veterinary services. County-level investments can also include the development of county legislation to devolve authority for the management of public goods such as the rangelands to recognised community-based organisations better placed to ensure their good governance, which is crucial for building adaptive capacity.

Integration of Climate Information into Adaptation Decision-Making

Appropriate, timely and targeted climate information is critical to enable county- and ward-level adaptation planning committees to prioritise investments that build adaptation and resilience to a changing climate. Climate information provided by the Kenya Meteorological Department (KMD) is integrated into the participatory vulnerability and resilience assessments or resilience assessments carried out by WAPCs to ensure that prioritised investments take into account current and future climate variability and hazards. County directors of meteorology (CDMs) play a key role in institutionalising climate information services (CIS) at county level. They are standing members of key decision-making fora such as the county adaptation planning committee responsible for ensuring WAPC proposals specifically addressing climate change and uncertainty. They are also responsible for the development of county CIS plans and facilitating the two-way communication of weather and climate information between the Meteorological Department and county and community actors. The CIS further provide an opportunity to invest in an observation network for more reliable and suitable data that can benefit all livelihoods, such as dryland crop farming and pastoralism. In addition various channels such as radio, SMS and extension officers are used to disseminate information at intervals (daily, weekly, monthly and seasonally), which has proved useful in ensuring that the information reaches the end users (community and households) (Ngari et al., 2016).

Monitoring and Evaluation of Adaptation

The Tracking Adaptation and Measuring Development (TAMD) monitoring and evaluation (M&E) framework is used to track how adaptation builds resilience and strengthens economic development (Brooks et al., 2013). TAMD as a framework has the advantage of being used in different contexts and scales to assess and compare the effectiveness of interventions that directly or indirectly assist populations in adapting to climate change (Karani et al., 2015). In Isiolo, the focus was on two fronts – first was helping communities through WAPCs develop their theories of change (ToCs) and relevant indicators for monitoring benefits from public goods investments; second, for county government, work focused on strengthening the existing systems to better support climate risk management. In the case of Isiolo, priority interventions included strengthening early warning systems, county budgeting and planning, and county coordination (Karani et al., 2015).

The WAPCs are the central 'pillars' of the approach piloted by the NDMA in seeking to mainstream climate into planning and democratise the allocation of climate finance for the benefit of the most vulnerable social groups (see Figure 15.1). Ensuring that the 'right' people become committee members is thus critical for the proper functioning of the CCCF. WAPC selection is thus often preceded by a comprehensive public information campaign to ensure that as many people as possible are aware and participate in the process of choosing their ward-level representatives. The vetting process,[4] as opposed to an election process which is often open to political manipulation and exclusion of the most marginalised, ensures that there is consensus on those selected to serve on the committee as well as a public commitment by those selected to be accountable to the wider community whose interests they represent. Communities put a lot of weight on integrity and leadership skills rather than academic qualification. The consortium thereafter provides training on more technical issues such as climate change, wider development-related policy and leadership skills, financial management and procurement to build people's skills as committee members. WAPCs often request support from county government technical officers and NGOs in the design of their projects (e.g. establishing budgets, M&E indicators) and their subsequent implementation.

The county adaptation planning committee on the other hand consists of a representative from each of the WAPCs, and representatives of national and county governments and NGOs. This ensures linkage with the lower level but also brings an additional set of skills and information on other planned activities by these institutions that may be relevant to the WAPCs.

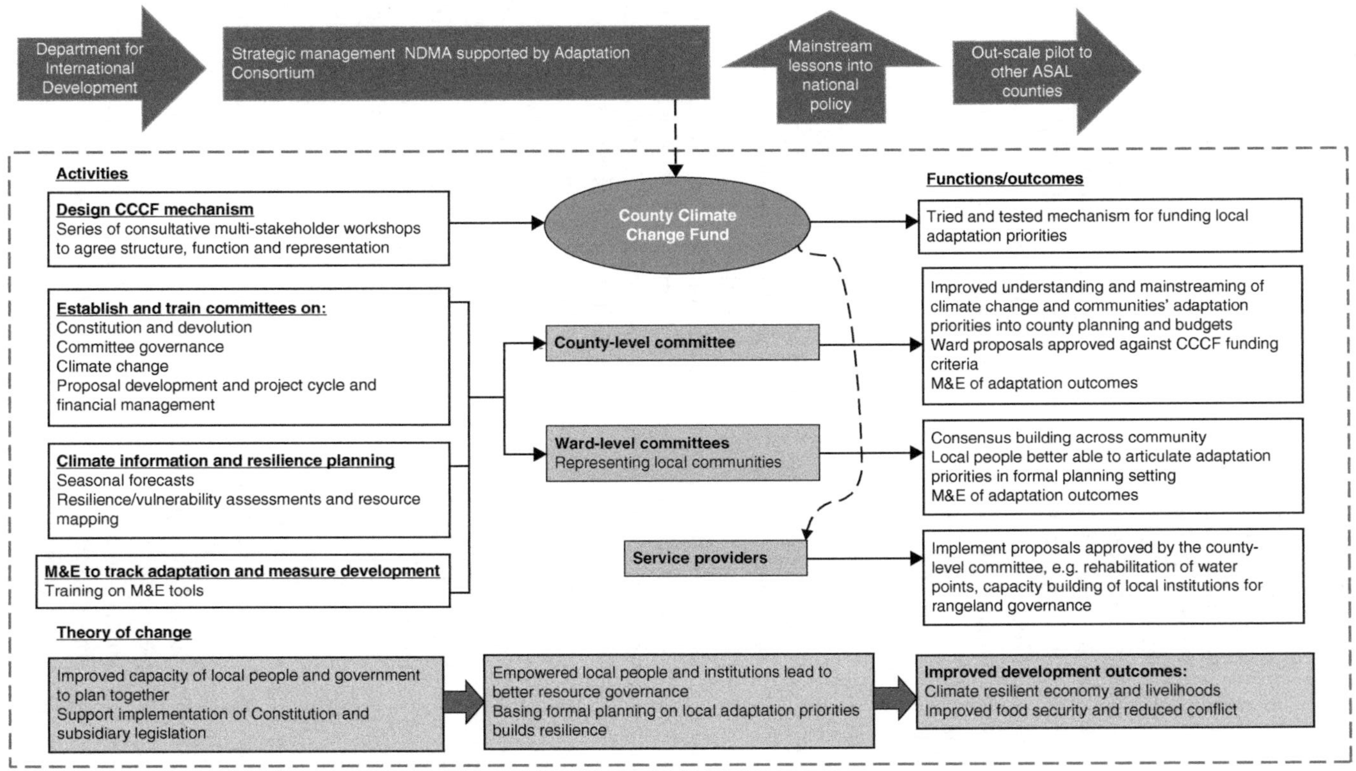

Figure 15.1 The County Climate Change Fund approach piloted by NDMA

Finally, fiduciary standards and safeguards, as set out in the CCCF procedures manual, are monitored through the following process:

- Payments of running costs and payments to service providers are conditional on demonstrating adherence to the steps and provisions in the procedure manual and constitutions.
- Auditors are engaged to conduct periodic spot checks of documentation and field visits to projects.
- CAPC meetings provide a forum for public accountability amongst WAPCs and the CAPC.[5]
- The CAPC, accompanied by the CCCF monitoring and evaluation officer from NDMA, conducts periodic monitoring visits to certify completion of projects, make technical recommendations on ongoing projects, and record adaptation impact against baseline indicators, as developed under the TAMD framework.

THE ISIOLO COUNTY CLIMATE CHANGE FUND: A SUCCESSFUL COMMUNITY-LED ADAPTATION INITIATIVE

Isiolo County was the first county to pilot the CCCF in 2012. The total cost for setting up the pilot Isiolo County Climate Change Fund (ICCCF) totalled £408 946 (NDMA, 2014a).

Since it was initiated in 2012, the ICCCF has completed two cycles of investments, funding a total of 42 projects costing £1 million over a three-year period (2012–2015). A total of 108 533 people from across the county have directly benefited from these investments.[6]

The investments are in five broad areas that address the underlying causes of vulnerability to climate change while strengthening adaptation to future extreme events:

- Mainstreaming of a climate change-related legal framework, by investing in establishing County Climate Change Fund legislation, a Livestock Strategy and Action Plan, and a Resource Atlas of Isiolo County.[7]
- Improved governance of the rangelands by funding activities of the *dedha* (customary range management institutions) to review and implement rules of access and control, including by pastoral groups who regularly visit Isiolo County, to be developed into county by-laws.
- Development of water infrastructure and training in water

governance in support of the multiple customary resource access rules and livestock mobility.
- Improved veterinary control through the rehabilitation of a decentralised livestock laboratory for disease surveillance and a county-wide vaccination programme as well as livestock handling facilities.
- Improved access to climate and other information by the development of a county CIS plan and strengthening the technical capacity of the community radio to broadcast to the whole county.

The first round of investment was largely focused on addressing the immediate development and adaptation deficits in Isiolo County in the face of current levels of climate variability and incidents of extreme weather events, for example improvements in rangeland governance to enable the *dedha* to better regulate the use of rainy and dry season pastures and drought reserves by both resident and visiting pastoralists with their livestock, thereby cushioning livestock from the effects of a difficult dry season or drought, or the deepening of surface water points to enable more water to be stored for a longer period.

The projects implemented in the second round were essentially as a result of gaps identified as well as lessons learned from the first round. For instance, those water points that are strategically located to facilitate the movement of livestock between dry and wet seasons were rehabilitated. This included in some cases closing certain water points that were inappropriately located and were contributing to over-stocking of livestock and social tensions between different pastoralist groups. In other areas, investments were made to reduce the waiting times for women at water points, thereby enabling them to spend more time on other activities that generate additional income that can be used to purchase food during the dry season without having to sell livestock, their main productive and reproductive assets. The continuous enforcement and monitoring of the rules established by the *dedha* contributed to pastoralists within Isiolo County managing the long dry period of 2015 without losing livestock.

Although it is too early to establish the sustainability of the achievements in Isiolo County, there is growing evidence that the two rounds of investments have had a significant impact on various aspects of pastoralism (Tari et al., 2015). For instance, during the dry season of 2014, a moment of great hardship for people and their animals in Isiolo County, the NDMA's official drought early warning system reported better than expected conditions for people and livestock in Isiolo compared to neighbouring counties despite similar rainfall conditions (NDMA, 2014b). The Isiolo CAPC attributed this in part to the improved governance of the rangelands funded by the ICCCF (NDMA, 2014a) (see Box 15.2).

BOX 15.2: TESTIMONY ON DRY SEASON RESILIENCE FROM BETTER GRAZING RESERVE MANAGEMENT

> Dedhas (Boran traditional resource management institutions) are mandated to regulate access to pasture and water in pastoral systems, yet the institutions continue to be weakened and undermined by formal system of governance. The support by Isiolo County Climate Change Fund came to strengthen customary systems of planning, use and management of our natural resources. The natural resource management meetings we have undertaken not only awakened our customary system of managing grazing land and water into the wet, dry and drought reserves, but also capacitated the [Dedha] members to do proper planning to enable effective use and utilization of resources. The planning process enabled the community to reclaim back community drought reserves; this move rubbed many up the wrong way, including political leaders who wanted to maintain the status quo at the expense of the majority. Communities have now put in place systems to regulate entry and access of pastoralists into these seasonal grazing areas. Our pasture land is now well managed and we have drought fall back areas. (Mzee Sar Goresa Dedha member Kinna, March 2014)

Source: NDMA (2014a).

Similarly, in 2015, there were no reported households that received food relief in those areas where there have been investments by the ICCCF (Tari et al., 2015). This can be partially attributed to the ICCCF investments, coupled with other factors such as good veterinary practices and the efforts of NDMA in drought management. A good example can be drawn from Oldonyiro Ward, where communities are able to access water throughout the year for both domestic and livestock uses. This has been possible as a result of the various water structures that were rehabilitated or created under the ICCCF, which are designed to complement each other in response to the opportunities offered by the local context. The communities accessed water initially from the rehabilitated rock catchment areas immediately after the rainy season, followed by the sand dams after the rock catchments had dried up, thereby providing them with water further into the dry season. The final water source used is the strategic borehole, which provided water throughout the year. The various sources of water helped people to reduce movement in accessing water and gave households more time to engage in other livelihood activities such as livestock marketing and sales of household consumables around the water points, which created an extra source of income.

The investments in Isiolo County have also created an opportunity for other counties to access shared resources. Pastoralists from Garissa, Wajir and Marsabit counties are among the beneficiaries of the ICCCF during the dry season and droughts. The ICCCF M&E field visit carried out in August 2015 found that the Hawaye shallow wells and Fororsa water pans located in Sericho Ward were accessed by pastoralists from the neighbouring counties of Wajir and Garissa during the prolonged dry spells of 2015.

In addition to material changes in local people's lives, the pilot is introducing more transformative changes. These have focused on institutional reforms to change the ways things are done, ensuring that the voices of the vulnerable are heard, and inform prioritisation and implementation of public goods investments. A CCCF bill has been enacted that institutionalises the fund as a public fund, ensuring synchronicity with the county annual planning and budgeting process. This provides the legal basis for counties to draw down climate finance from national and international sources (e.g. the GCF). Furthermore, the Isiolo CCCF bill commits the county government to contribute a minimum of 2% of its development budget to climate finance, and empowers local people to prioritise 70% of available climate finance. This ensures that the approach is context driven, inclusive and under the oversight of the local community – all of which will contribute to making devolution work. The county CIS plan will strengthen further mainstreaming efforts, which started with the review of the County Integrated Development Plan (CIDP) and development of an integrated livestock development strategy.

Evidence suggests that the ICCCF is having an impact on county government and donors. Activities in March 2014 saw increased technical input from county government technical officers in the second round, and an ICCCF-funded community consultation process for both the County Integrated Development Plan – a key document for county planning administration – and the County Livestock Strategy. More recently, the ICCCF has supported the design of a multi-sectoral Strategic Plan for Water and Irrigation, Energy, Environment and Natural Resources. Additionally, county government has provided in-kind support for the veterinary lab and vaccination campaign, and KMD is continuing to support climate information services through the provision and training of staff, and the deployment of climate observation equipment, as well as managing a community radio station that is used to discuss various development and climate-related issues. Representatives in the Isiolo County Assembly have also advocated for the ICCCF process. This, coupled with indications of further funding support from county and development partners, demonstrates the broader potential impact of the ICCCF process.

LESSONS LEARNED

A key lesson to emerge from the work in Isiolo County is that it is necessary to first address the 'development deficit', common to many arid and semi-arid lands (ASAL) in Kenya and Africa more generally, before it is possible to engage local communities in considering longer-term climate adaptation. Climate change exacerbates existing structural causes of poverty and inequality, and in the dryland areas of Kenya a historical legacy of limited and often inappropriate development has left these areas with weaker institutions of governance and planning, less effective social and economic services, and greater levels of poverty than other areas of the country. This increases their vulnerability to current climate variability and extreme events.

The success of the approach can be attributed to stronger involvement of the technical officers at the county level, who often have competing demands on their time. Mainstreaming such approaches therefore requires patience and working at the officers' pace rather than project timelines.

The work in Isiolo has highlighted the common difficulties faced by communities in underdeveloped dryland areas, including:

- A lack of technical experts to support development activities. For example, in Isiolo there is one water officer supporting the design and supervision of several water projects.
- A generalised lack of understanding of the rationale of dryland economies and livelihood strategies amongst many government staff, resulting in them seeking to introduce or impose often inappropriate approaches or technologies that inadvertently further increase local people's exposure to climate risk.
- A lack of general communication, transport and other infrastructure, thus reducing the efficiency and speed of project activities. This has partly contributed to lack of service providers to implement work at the local level, thus making the procurement process difficult.
- A lack of general banking and accounting infrastructure and practice. For example, all transactions require supporting documents to be submitted and filed for auditing, but there is a general difficulty of getting receipts.
- Weather and seasonal variability can interfere with project implementation and monitoring plans, thus making the process difficult to align with the fixed financial calendars of government or donors.

CONCLUSION

The continued success of the ICCCF is serving to inform the application of the approach in the four other Kenya ASAL counties (Wajir, Garissa, Kitui and Makueni) with support from the Adaptation Consortium. Given the differing contexts between counties, and that county government structures are generally more developed now than at the inception of the ICCCF, the structure and process of the CCCF approach will likely vary across the four new counties, but the ICCCF provides a basis for the design of these approaches. In this respect, the CCCF approach is of equal relevance to all county governments in Kenya. The ICCCF experience has led to three other African countries adopting the same approach: Tanzania, Mali and Senegal.[8]

The approach being piloted by the NDMA through the Adaptation Consortium shifts away from current practices by institutionalising a more accountable manner in which (climate) finance is managed at the local government level, creating an environment where governments and community representatives engage with each other on a more equal footing for sustainable adaptation planning. In empowering local people, the approach also contributes to the identification of more appropriate responses to the unpredictable and increasingly variable conditions likely to occur as a result of climate change. In these conditions, it is essential that decision-making and resources be managed by inclusive local institutions that have capacity and flexibility to manage risks through flexible 'real-time' adaptation, avoiding the bad or worst consequences while taking advantage of the opportunities that can also occur. This community-driven approach offers greater potential to tackle marginalisation and inequality issues that increase the vulnerability of women and other excluded groups.

Institutionalising this approach within government planning and budgeting will contribute to sustaining a process of empowering county governments and communities for a more transformational adaptive approach to addressing climate change.

NOTES

1. Broadly, climate finance may be defined as money mobilised to finance activities that reduce greenhouse gases emission and/or support adaptation to the impacts of climate change. It may be broadly classified as either public, when it comes from government revenue, or private if it comes from the private sector, which may range from individuals, small businesses and corporations to charities among others (UNFCCC, 2014; Rai et al., 2015).
2. Kenya's Constitution of 2010 introduced devolved government consisting of county-, sub-county- and ward-level political and administrative units responsible for planning

and service delivery. The wards are the lowest planning level and thus closest to local people.
3. The Adaptation Consortium brings together Anglican Development Services–Eastern (ADS-E), Arid Lands Development Focus (ALDEF), Christian Aid, International Institute for Environment and Development (IIED), Kenya Meteorological Department (KMD), Resource Advocacy Programme (RAP), UK MET and Womankind Kenya. The Adaptation Consortium is a core component of the Department for International Development (DFID)-funded Strengthening Adaptation and Resilience to Climate Change in Kenya Plus programme (2013–2017).
4. The vetting process consists of inviting those identified for leadership positions to step outside and allow the larger community to discuss and agree on the most suitable.
5. The community radio in Garba Tulla, once fully operational, is expected to play a major role in increasing the accountability of the WAPC members to the local community, and for community monitoring of the implementation of projects, with committees being required to receive public feedback on air.
6. The number of beneficiaries is established through the collection of data from triangulated sources: direct counting by the WAPCs' management committee responsible for monitoring and the county-level adaptation committee during regular M&E visits – both supported by their respective share of the 10% administration fund of the CCCF.
7. Additional strategies and legislation in preparation include: the Isiolo County Strategic Plan for Water Energy and Environment Sector and the Isiolo County Rangeland Management Bill.
8. See: http://www.iied.org/responding-climate-change-east-africa-strengthening-dryland-governance-planning; and http://www.iied.org/decentralising-climate-funds-mali-senegal.

REFERENCES

Brooks, N., S. Anderson, I. Burton, S. Fisher, N. Rai and I. Tellam (2013), *An Operational Framework for Tracking Adaptation and Measuring Development*, Climate Change Working Paper No. 5, London: International Institute for Environmental Development (IIED), http://pubs.iied.org/pdfs/10038IIED.pdf.

Christensen, K., S. Raihan, R. Ahsan, A.M. Nasiruddin, S. Ahmed and H. Wright (2012), *Financing Local Adaptation: Ensuring Access for the Climate Vulnerable in Bangladesh*, Dhaka: ActionAid Bangladesh, Action Research for Community Adaptation in Bangladesh, Bangladesh Centre for Advanced Studies, and International Centre for Climate Change and Development, http://www.actionaid.org/sites/files/actionaid/financing_local_adaptation.pdf.

Gupta, S., J. Harnisch, D. Barua, L. Chingambo, P. Frankel, R. Garrido Vasquez, L. Gomez-Echeverri, E. Haites et al. (2014), 'Cross-cutting investment and finance issues', in O. Edenhofer, R. Pichs-Madruga, Y. Sokona, J.C. Minx, E. Farahani, S. Kadner, K. Seyboth, A. Adler et al. (eds), *Climate Change 2014: Mitigating Climate Change: Contribution of Working Group III to Fifth Assessment Report of the Intergovernmental Panel on Climate Change*, Cambridge, UK: Cambridge University Press, pp. 1207–1246.

Karani, I., J. Mayhew and S. Anderson (2015), 'Tracking adaptation and measuring development in Isiolo county, Kenya', in D. Bours, C. McGinn and P. Pringle (eds), *Monitoring and Evaluation of Climate Change Adaptation: A Review of the Landscape*, New Directions for Evaluation 147, San Francisco, CA: Jossey-Bass, pp. 75–87.

Mapesa, B. and T.N. Kibua (2008), 'Management and utilization of Constituencies

Development Fund in Kenya', in T.N. Kibua and G. Mwabu (eds), *Decentralisation and Devolution in Kenya: New Approaches*, Nairobi: University of Nairobi Press, pp. 236–251.
NDMA (2014a), 'Isiolo County Adaptation Fund: Activities, costs and impacts after the 1st investment round', Project report, Ada Consortium and NDMA.
NDMA (2014b), *Early Warning Bulletin, Isiolo, July, 2014*, Nairobi: National Drought Management Authority.
Ngari, Francis Maina, G.A. Obuya, X. Hou, G. Larson and A. Braimoh (2016), *Climate Information Services Providers in Kenya*, Agriculture Global Practice Technical Assistance Paper, Washington, DC: World Bank Group, http://documents.worldbank.org/curated/en/2016/02/25961964/climate-information-services-providers-kenya.
Rai, N., N. Kaur, B. Wang and P. Steele (2015), *Topic Guide: A Guide to National Governance of Climate Finance*, London: Evidence on Demand, http://dx.doi.org/10.12774/eod_tg.march2015.rainetal.
Steele, P., N. Rai, A. Walnycki and I. Nhantumbo (2015), *Delivering Climate and Development Finance to the Poorest: Intermediaries that 'Leave No-One Behind'*, IIED Briefing Paper, November, London: International Institute for Environment and Development, http://pubs.iied.org/pdfs/17317IIED.pdf.
Tari, D., C. King-Okumu and I. Jarso (2015), 'Strengthening local customary institutions: A case study in Isiolo County, Northern Kenya', Ada Consortium, NDMA, http://www.adaconsortium.org/images/publications/Rapid_Assessment_-_Web_Update.pdf.
UNFCCC (2014), *UNFCCC Standing Committee on Finance: 2014 Biennial Assessment and Overview of Climate Finance Flows Report*, Bonn: UNFCCC.
UNFCCC (2015), Adoption of the Paris Agreement, Conference of the Parties, Twenty-first session, https://unfccc.int/resource/docs/2015/cop21/eng/l09r01.pdf.

16. Do climate policies hurt the economy? Lessons from the EU experience

Carlo Carraro and Marinella Davide

INTRODUCTION

From the 1980s, the first measures to mitigate greenhouse gases (GHGs) started to be implemented in Europe, fostered also by the growing scientific and political debate at the international level. In 1996 the European Union (EU) became the first political body to support the long-term stabilization target – the 2°C limit – which soon after emerged as a reference point for global climate action (European Union Council, 1996). At a similar time, under the Kyoto Protocol, the EU committed itself to cutting GHG emissions by 8%, compared to 1990 levels, between 2008 and 2012. To achieve this target, the Union launched a broad range of policies, including the world's first regional Emissions Trading System (EU ETS) in 2005 (EC, 2003).

Four years later, Brussels adopted an unprecedented set of measures: the 2020 Climate and Energy package, which mandates an emission reduction of at least 20% by 2020, in relation to 1990, combined with a 20% increase in renewable energy sources, and an indicative 20% of energy efficiency improvements (EC, 2008). As part of a broader long-term decarbonization vision, in 2011 the European Commission (EC) established a roadmap to reduce emissions by 80–95% by 2050, in relation to 1990. With this objective in mind, in 2015 a new low-carbon framework for 2030 was defined, aimed at cutting GHGs by 40%, in relation to 1990, by 2030, increasing the renewable share to 27% of energy consumption, and reaching at least 27% of energy savings compared to the business-as-usual scenario (EC, 2014). This target is also at the base of the EU contribution proposed in the context of the Paris Agreement, the landmark climate deal reached within the UNFCCC at the end of 2015 (UNFCCC, 2015).

The process of building and consolidating a European climate change strategy is the result of decades of efforts and compromises both inside and outside the EU. Among the concerns that have accompanied political choices over the years, those related to the costs and the economic impacts of mitigation measures have been the most debated.

Against this background, the research community has made substantial efforts to periodically assess EU climate interventions. Building on the recent literature, the chapter reviews the studies analyzing the economic impacts of the EU mitigation policies, including the macroeconomic costs, the impacts on competitiveness and international trade, and the potential distributive effects.

Evaluating the European climate policy is an ambitious task, since it includes very different measures. For this reason our review focuses mainly on the EU 2020 emissions reduction target, the ETS as a pillar of the EU's climate policy, and, where a national perspective is needed, the environmental tax reforms (ETR).

THE ECONOMIC IMPACTS OF EU MITIGATION POLICIES

Climate change mitigation policies aim primarily at limiting the release of GHGs into the atmosphere in order to reduce negative environmental externalities. In addition, as recently stressed also by the IPCC (2014), mitigation is a key component of the so-called "climate-resilient pathways", which, along with adaptation policies, can help to reduce the risks of climate change to natural and human systems and promote sustainable development.

Economic theory usually groups policy instruments for reducing emissions into two categories: command-and-control regulation (including emission quotas, and pollution or technology standards) and market-based instruments (carbon taxes, subsidies, and cap-and-trade systems). The latter have usually been considered more flexible and efficient than traditional prescriptive regulation, as they encourage firms and individuals to undertake pollution reduction through market signals (Freeman and Kolstad, 2007).

Whatever the instrument, the traditional perspective suggests that abatement of emissions imposes, directly or not, additional costs on the economy.

The literature simulating economic impacts of mitigation policies is vast and it mainly includes modelling studies, using integrated assessment models (IAMs), which project different types of cost related to environmental measures. Unfortunately, the direct benefits of mitigation through avoided impacts have received little attention so far. Despite increased research efforts, the lack of reliable metrics for estimating the benefits of mitigation, especially in monetary terms, prevents them from being compared to the costs (see Corfee-Morlot and Agrawala, 2004). This is why the issue is not addressed in the following sections.

The Effect on GDP and Welfare

One of the main cost metrics used to evaluate mitigation policy is its economy-wide costs, which are usually reported as changes in GDP, welfare, consumption or energy system costs (Paltsev and Capros, 2013). When abatement measures lead to relocation of investment from more productive to less productive uses, there is a loss of productivity, representing a net cost for the economy.

As for the EU's welfare losses or GDP costs, the studies under review agree that they would be rather limited, especially if full flexibility across sectors were allowed (Table 16.1). Indeed, the decision to split the carbon

Table 16.1 Cost of the EU emissions reduction policy

Reference	Model (type)	Policy	Rate
Peterson, S. and Klepper (2008)	DART (CGE*)	−20% target	0.2–1.6% GDP loss in 2020.
Böhringer et al. (2009)	DART, PACE, Gemini-E3 (CGE)	−20% target	0.5–2.0% GDP loss in 2013–2020. Fragmentation increases the cost by 50–90%.
EC (2010)	GEM-3, PIL PRIMES (bottom-up energy)	20-20-20 (post-crisis)	0.32% GDP loss (global EU-27).
Peterson, E. et al. (2011)	DYE-CLIP (CGE)	−20% target	Increase in GDP ≈ +0.1%.
Alexeeva-Talebi et al. (2012)	(CGE)	Emission reduction target (up to −30%)	1.6–2.5% reduction in real consumption (20–30%).
Bosello et al. (2013)	ICES (CGE)	20-20-20 package	0.56% GDP loss in 2020.

Note: * Computable general equilibrium (CGE) models are a widely used tool for the economic analysis of environmental externalities and mitigation policies (see Sue Wing, 2011 for an introduction).

price across ETS and non-ETS sectors introduces some inefficiencies that drive compliance costs up. According to S. Peterson and Klepper (2008), the EU welfare losses in 2020 could be low (0.2%) under an efficient and all-encompassing market, or even if the use of international credits (e.g. credits under the Kyoto Protocol's Clean Developed Mechanisms) is not allowed (0.5%). However, the aggregate loss rises to 1.6% if trade is limited to ETS sectors. Among member states, losses vary, with Eastern Europe actually projected to gain from the ETS because of its excess emission permits.

Similarly, Böhringer et al. (2009) add that abatement cost grows by 90% if a specific target is placed on renewables. Opposite results are drawn from E. Peterson et al. (2011), who, in the case of an international action such as the Copenhagen Accord's pledges, find that the EU's GDP increases because of the relative lower energy and carbon intensities that make EU firms more competitive.

The Impact on Firm-Level Competitiveness

Competitiveness is a key economic indicator to monitor when an environmental policy measure is implemented. Despite the many possible interpretations, competitiveness can be defined as the measure of a country's advantage or disadvantage in selling its products in international markets.[1] An emissions reduction policy can indeed affect the competitiveness of countries through different channels: it may require firms to improve their environmental or energy efficiency performances, thus increasing production costs, or, as in the case of ETS, it directly requires firms to buy allowances to cover their emissions. In some cases, a mix of the two effects can be experienced (see OECD, 2003 for an overview).

It is important to distinguish between competitiveness effects at the country, sectoral and firm levels (Ekins and Speck, 1999). Emission reduction measures may lead some sectors to shrink while others can expand because of a relocation of production factors and resources. Therefore, the country performance can differ from the firm or sectoral performance.

However, according to the Porter hypothesis, a well-designed environmental regulation may foster innovation, productivity and competitiveness, especially in the long run (Porter and van der Linde, 1995). Since the Porter hypothesis considers "excessive pollution" as the signal of an inefficient use of resources, interventions aimed at correcting it could move the production system closer to the efficiency frontier, with net monetary gains and incentive towards technological innovation.

In addition, policy design can play a non-negligible role in the distribution of costs and benefits across sectors. Particularly, allocating emission

Table 16.2 Effects of the EU's mitigation policy on competitiveness (EU ETS versus business-as-usual scenario and EU ETS versus no-trade scenario)

Reference	Model (type)	Reference scenario	Effects on competitiveness
COWI (2004)	GTAP-ECAT (CGE)	BAU	Negative effect: output −0.36% (−0.48% with sluggish technology adaptation).
Klepper and Peterson (2004)	DART (CGE)	BAU	Negative effect: overall output −0.3%, energy sector's output −2%.
Reinaud (2008)	(Partial equilibrium model)	BAU	Most sectors: very small and diverse effects. Negative effect: aluminum industry, costs +3.7%, demand −2.9%.
Smale et al. (2006)	(Cournot oligopoly model)	BAU	Positive effect on earnings: cement, printing, petroleum, steel. Negative effect: aluminum industry, −100% earnings.
Alexeeva-Talebi et al. (2012)	(CGE)	BAU	Negative effect: energy-intensive sectors' output −6%, mineral oil industry output −25%.
Bosello et al. (2013)	ICES (CGE)	BAU	Negative effects on energy intensive sectors' output: −1.5%; non-ETS sectors: −0.3%; ETS sectors: −2% (excluding electricity).
Klepper and Peterson (2004)	DART (CGE)	No trade	Positive effect on overall small output growth, energy sectors' output +3%.
Böhringer et al. (2005)	SIMAC (partial equilibrium model)	No trade	Negative effects: compliance costs +400% (actual national allocation plans, costs accrue mainly in non-participating sectors).
Kemfert et al. (2005)	GTAP-E (CGE)	No trade	Positive effects: abatement costs −98% (maximum).

permits free of charge (grandfathering) is typically invoked as a means of reducing compliance costs and competitive losses or even subsidizing the regulated sectors. This is the case for the EU ETS, where allowances were given free of charge in the initial phases (2005–2012) before gradually shifting toward an auctioning system, with the exception of sectors risking carbon leakage.

Moreover, revenues from carbon-energy taxes offer the possibility of shifting the burden of taxation away from capital or labor towards pollution to increase growth and/or employment, in addition to the environmental benefit. Potential for this "double" or even "triple" dividend has fostered a huge applied and theoretical debate (Jaeger, 2012).

Overall, the analysis of recent simulation studies on EU policy concludes that emissions reduction leads to some negative effects on competitiveness, especially for energy-intensive sectors (Table 16.2). In most cases, the estimated effect is rather modest, amounting to an overall output contraction of 0.3% compared to a no-policy scenario (COWI, 2004; Klepper and Peterson, 2004). One exception is Smale et al. (2006), who find a positive impact on four out of the five sectors analyzed. According to the author, the grandfathered allowances more than offset the marginal cost growth implied by the ETS. Aluminum results as the most affected sector (Smale et al., 2006; Reinaud, 2008).

Alexeeva-Talebi et al. (2012) suggest that a moderate tax differentiation would sectorally balance the competitiveness effects of mitigation policies and limit overall efficiency losses.

Once again, the role of trading needs to be highlighted: a well-designed ETS can reduce costs (between 25% and 98%) or lead to competitiveness gains (Oberndorfer and Rennings, 2007). This is confirmed by studies that compare the competitiveness effects of the EU's mitigation target with ETS against a scenario without trade (Table 16.2).

In contrast with this view is the analysis by Böhringer et al. (2005), according to which restrictions on the EU's ETS induce additional costs and consequent losses in competitiveness. The results derive from the assumption that the generous initial allocation of allowances shifts abatement obligations from ETS to non-ETS sectors. This conclusion agrees, however, with the others in suggesting that a broader implementation of the EU's ETS would improve competitiveness.

The Carbon Leakage Effect

When analyzing competitiveness issues, it is also important to consider the international dimension. The "carbon leakage" effect denotes an increase in GHG emissions in countries not subject to strict emission regulation, as a reaction to the constraint imposed on a country or a group of countries. This may happen when carbon-intensive industries relocate production to non-regulated countries to continue producing at lower costs (pollution haven) or because consumer demand shifts from expensive "clean" goods towards "dirtier" and cheaper substitutes produced in non-regulated countries (see Reinaud, 2008).

Beyond the undermined effectiveness of emission reduction efforts, carbon leakage reveals a policy-induced loss of international competitiveness of regulated carbon-intensive sectors. In this simplistic scheme, mitigation policies entail costs for implementing countries, and competitive gains for non-regulating countries. To offset potential competitive disadvantages, countries may, in theory, implement the so-called border tax adjustments (BTAs), consisting of taxes based upon the carbon content of foreign commodities, aimed at increasing the cost of and reducing the demand for imported energy-intensive products.

The recent literature on carbon leakage highlights the great variability of ex-ante modelling analyses and the difficulty of drawing firm conclusions

Table 16.3 Carbon leakage effect as a consequence of the EU's ETS

Reference	Model (type)	Policy assumptions	Carbon leakage ratio*
Ponssard and Walker (2008)	Cement trade and competition model (oligopoly competition)	€30–€50/tCO_2 carbon price, 100% auctioning, no BTA, focus on cement sector	65–73%
Kuik and Hofkes (2010)	GTAP-E (CGE)	€20/tCO_2	10.8%
Monjon and Quirion (2011)	CASE II (partial equilibrium model)	€24 carbon price, 100% auctioning, no BTA, focus on four sectors: electricity, steel, cement and aluminum	4.5–11.4% (low–high elasticity) aggregated effect in 2016
Alexeeva-Talebi et al. (2012)	(CGE model)	EU emission reduction target (20%), no BTA	27%
Branger and Quirion (2014)	Meta-analysis	N/A	3% to 33% (mean 19%)

Note: * Computed as the emission variation in the rest of the world over the emission variation in the climate coalition (difference between emissions in the climate policy scenario and the counterfactual business-as-usual scenario).

because of persistent doubts about the robustness of the results depending on different assumptions (Table 16.3). Insights into the sectoral characterization of leakage are provided by the survey from the Öko-Institut (2013): it seems to be high for the cement sector (19% to 73%), even though decreasing (50%) with grandfathering and output-based allowances allocation (9%). Estimates for the iron and steel sector range between 35% and 40%, while the aluminum sector's leakage rates are at 20% (Öko-Institut, 2013).

A meta-analysis on 310 estimates conducted by Branger and Quirion (2014) shows that the carbon leakage ratio is on average 19%.

The implementation of border tax adjustments is projected to limit the rate of carbon leakage from 15% ("light" BTA) to 9% ("strong" BTA) (Branger and Quirion, 2014). However, other studies point out that the overall impacts and benefits are not large enough to justify implementation of BTAs, given their administrative complexity and the deleterious effects on international trade (McKibbin and Wilcoxen, 2008).

The Distributional Effects of the EU Climate Policy

To complete the overview of the economic impacts of emission abatement interventions in the EU, this subsection focuses on the distributional effects among households.

Economic theory identifies three major effects. The first direct effect is the price increase of energy products purchased by households (e.g. fuel for heating or gasoline for autos). The second effect is indirect, defined as the pass-through of this price increase in all downstream goods and services. The third effect includes impacts on growth and the labor market (Grainger and Kolstad, 2009). Eventually, all these impacts may lead to important changes across households' income distribution.

Recent research on climate-related distributional effects has focused almost exclusively on domestic ETRs, mainly because cap-and-trade schemes have emerged only recently in comparison with taxation mechanisms.

The general pattern expressed by analyses in EU countries is that energy and carbon taxes are overall mildly regressive,[2] but their impacts on low-income households are less than first expected. As expected, the recycling of the revenue generated is one of the major drivers of distributional effects. Moreover, the behavioral response, through product substitution possibilities and price elasticities, can alleviate the level of regressivity.

Early simulations of Germany's ETR finds that lower-income households, especially those dependent on transfers (pensioners and the unemployed), bear a heavier burden as a share of net household income, mainly because of higher prices for electricity, heating oil and natural gas (Bach

Table 16.4 Distributional effect of the EU's emissions reduction policy

Reference	Model (type)	Policy	Distributional effects
Bach et al. (2002)	(Potsdam micro-simulation model)	Energy taxes in Germany	Regressive: small magnitude (maximum effect on low-income −1.4%).
Bork (2006)	(Micro-simulation model)	Additional excise duties on energy products in Germany	Weak regressivity. Average loss −0.7% of income.
Dresner and Ekins (2006)	POLIMOD (microsimulation tax-benefit model)	Carbon tax (£10/tCO_2) in the UK	Burden of 0.51% of annual income for the first decile, 0.12% for the tenth decile.
Kerkhof et al. (2008)	Input−output analysis (IOA)	Carbon and multiple GHG taxes in the Netherlands	Regressive: first decile pays 7% of income, while tenth decile pays 4%.
Callan et al. (2009)	SWITCH (microsimulation tax-benefit model)	Carbon tax (€20/tCO_2) in Ireland	Less than €3/week to poorest households, more than €4/week to richest.
EEA (2011)	E3ME (global macroeconometric model)	Energy tax in EU-27	Positive change in real incomes for all socio-economic groups. Larger benefits for the employed and urban residents. The third quintile gains the least.

et al., 2002). Similar results are driven by Bork (2006), who estimates an average aggregate income loss of about 0.7%. Among the taxes considered by the German reform, the only one that does not show a regressive effect of up to the €25 000 to €30 000 annual income bracket is the motor fuel tax. When revenues are recycled in the form of a reduction in social security contributions, the regressive effect is strengthened, even though remaining weak (1% of income). Similar findings are driven by Dresner and Ekins (2006), who also model 13 different compensating packages and find they offset the result for the lowest deciles (gains from £1.77 to £118.14 per year). However, a minimum of a fifth of the poorest households remain losers (even assuming full take-up of benefits).

An analysis by the European Environment Agency focuses on a Europe-wide ETR (EEA, 2011) and comes to opposite conclusions. A tax on energy and material inputs whose revenues are recycled through labor taxation would create a positive change in real incomes for all socio-economic groups. However, groups that do not pay income taxes (unemployed, retirees, etc.) see a smaller increase, as they do not benefit from the tax reduction (Table 16.4).

Including indirect effects in the distributional analysis appears to decrease the overall regressive impact: households with lower incomes have a smaller consumption basket than those with higher incomes, and are thus less impacted by the global rise of goods and services due to climate policy. Likewise, Boyd et al. (1995) explain that energy taxes will cause a drop in the price of return of capital, which will hurt to a relatively larger extent the upper income groups (since they derive a larger fraction of their income from capital ownership).

CONCLUSIONS

The recent efforts carried out by the research community are certainly helpful to better understand the economic implications of the EU's climate policy. Overall, the ex-ante modelling analysis shows that the worries about costs induced by climate regulation are usually overestimated.

Welfare losses as a consequence of the EU's mitigation target have been found to be limited, even though policy design is crucial in limiting them. Specifically, fragmented policies or overlapping objectives – such as the European differentiation between ETS and non-ETS sectors or the single target for renewables – lead to inefficiencies, thus increasing compliance costs as compared to the implementation of a carbon price encompassing all sectors.

As for competitiveness, emission reduction policies can reduce productivity and employment by small amounts, in particular in energy-intensive sectors, and especially during the transitory period. Especially if cap-and-trade mechanisms are in place, as in the case of the EU's ETS, the effect on the competitiveness of regulated firms and sectors can be negligible. Carbon leakage effects are more difficult to estimate, because of the strong dependence on modelling assumptions. However, moving from theory to practice, the recent experience with the EU's ETS highlights that the structure of energy-intensive sectors and international dynamics other than climate regulation (e.g. energy prices and international demand) are more relevant to the decisions of firms to relocate (Sartor, 2012; Bolscher et al., 2013; Dechezleprêtre et al., 2015). Moreover, the threat of carbon leakage is constantly declining because of the increase in measures adopted by other jurisdictions, such as the trading systems in China and the huge number of country contributions submitted in the context of the UNFCCC climate deal reached in Paris in December 2015.

On the other hand, recent studies on the historical data of ETS highlight that both costs and impacts have been limited also because of the low price of allowances caused by the abundant allocation of free permits in the first periods of the scheme (Ellerman et al., 2010; Dechezleprêtre and Sato, 2014). In the third period (2013–2020), it is likely that that situation will change as the auctioning of allowances becomes the rule and market reforms will be implemented to make the cap more stringent.

As for distributional effects, the overall effect seems harder to pinpoint: the apparent regressivity is reduced through the use of behavioral response and the inclusion of indirect effects. Also, revenue recycling from carbon-energy taxes offers the possibility of shifting the burden of taxation away from labor towards pollution.

Finally, benefits from improvements in environmental quality are almost always disregarded because of the extreme complexity of assessing them. Trying to better perform cost and benefits analysis of mitigation policies, including environmental or health benefits, will certainly help to get a broader view of the real effect of climate change policies.

ACKNOWLEDGEMENTS

This research was supported by the CliMiP (Climate Change Mitigation and Poverty Reduction – Trade-Offs or Win–Win Situations?) project funded by the Volkswagen Foundation under its "Europe and Global Challenges" program.

NOTES

1. OECD's glossary of statistical terms, https://stats.oecd.org/glossary/detail.asp?ID=399 (accessed 20 May 2016).
2. In the literature, "regressive" means that the poorest households bear the brunt, "progressive" that the richer ones are in this position, and "neutral" that all households are similarly affected (in relative terms).

REFERENCES

Alexeeva-Talebi, V., C. Böhringer and U. Moslener (2012), *Climate and Competitiveness: An Economic Impact Assessment of EU Leadership in Emission Control Policies*, ZEW Technical Report, Mannheim, Germany: Centre for European Economic Research (ZEW).

Bach, S., M. Kohlhaas, B. Meyer, B. Praetorius and H. Welsch (2002), 'The effects of environmental fiscal reform in Germany: A simulation study', *Energy Policy*, **30**, 803–811.

Böhringer, C., T. Hoffmann, A. Lange, A. Löschel and U. Moslener (2005), 'Assessing emissions regulation in Europe: An interactive simulation approach', *Energy Journal*, **26**, 1–22.

Böhringer, C., T.F. Rutherford and R.S.J. Tol (2009), 'The EU 20/20/2020 targets: An overview of the EMF22 assessment', *Energy Economics*, **31**, S268–S273.

Bolscher, H., V. Graichen, G. Hay, S. Healy, J. Lenstra, L. Meindert, D. Regeczi, M. von Schickfus et al. (2013), *Carbon Leakage Evidence Project – Factsheets for Selected Sectors*, Rotterdam, the Netherlands: Ecorys.

Bork, C. (2006), 'Distributional effects of the ecological tax reform in Germany: An evaluation with a microsimulation model', in Y. Serret and N. Johnstone (eds), *The Distributional Effects of Environmental Policy*, Cheltenham, UK and Northampton, MA, USA: Edward Elgar Publishing In Association with the OECD, pp. 139–170.

Bosello, F., L. Campagnolo, C. Carraro, F. Eboli, R. Parrado and E. Portale (2013), *Macroeconomic Impacts of the EU 30% GHG Mitigation Target, 2013*, FEEM Working Paper 028, Venice, Italy: Fondazione Eni Enrico Mattei (FEEM).

Boyd, R., K. Krutilla and W. Kip Viscusi (1995), 'Energy taxation as a policy instrument to reduce CO_2 emissions: A net benefit analysis', *Journal of Environmental Economics and Management*, **29**, 1–24.

Branger, F. and P. Quirion (2014), 'Would border carbon adjustments prevent carbon leakage and heavy industry competitiveness losses?', *Ecological Economics*, **99**, 29–39.

Callan, T., S. Lyons, S. Scott, R. Tol and S. Verde (2009), 'The distributional implications of a carbon tax in Ireland', *Energy Policy*, **37**, 407–412.

Corfee-Morlot, J. and S. Agrawala (2004), The benefits of climate policy [Special issue], *Global Environmental Change*, **14** (3), 195–298.

COWI (2004), *Competitiveness and EU Climate Change Policy*, Report, Copenhagen: UNICE.

Dechezleprêtre, A. and M. Sato (2014), *The Impacts of Environmental Regulations on Competitiveness*, Policy brief, November, London: Grantham Research Institute on Climate Change.

Dechezleprêtre, A., C. Gennaioli, R. Martin, M. Muûls and T. Stoerk (2015), *Searching for Carbon Leaks in Multinational Companies*, Working Paper No. 187, Leeds, UK: Centre for Climate Change Economics and Policy.

Dresner, S. and P. Ekins (2006), 'Economic instruments to improve UK home energy efficiency without negative social impacts', *Fiscal Studies*, **27**, 47–74.

EC (European Commission) (2003), *Directive 2003/87/EC of the European Parliament and of the Council establishing a scheme for greenhouse gas emission allowance trading within the Community and amending Council Directive 96/61/EC*, OJ L 275.

EC (European Commission) (2008), *Communication from the Commission to the European Parliament, the Council, the European Economic and Social Committee and the Committee of the Regions 20 20 by 2020 – Europe's climate change opportunity*, COM(2008)30 final, Brussels.
EC (European Commission) (2010), 'Analysis of options to move beyond 20% greenhouse gas emission reductions and assessing the risk of carbon leakage', Commission Staff working document.
EC (European Commission) (2014), *Communication from the Commission to the European Parliament, the Council, the European Economic and Social Committee and the Committee of the Regions, 'A policy framework for climate and energy in the period'*.
EEA (European Environment Agency) (2011), *Environmental Tax Reform in Europe: Implications for Income Distribution*, Technical Report 16/2011, Copenhagen: EEA.
Ekins, P. and S. Speck (1999), 'The impact of environmental policies on competitiveness, theory and evidence', in T. Barker and J. Kohler (eds), *International Competitiveness and Environmental Policies*, Cheltenham, UK and Northampton, MA, USA: Edward Elgar Publishing, pp. 33–70.
Ellerman, A.D., F.J. Convery and C. Perthuis (2010), *Pricing Carbon: The European Union Emissions Trading Scheme*, Cambridge, UK: Cambridge University Press.
European Union Council (1996), 'Community strategy on climate change – Council conclusions', in *939th Council Meeting Environment*, No. 8518/96.
Freeman, J. and C.D. Kolstad (eds) (2007), *Moving to Markets in Environmental Regulation: Lessons from Twenty Years of Experience*, New York: Oxford University Press.
Grainger, C.A. and C. Kolstad (2009), *Who Pays a Price on Carbon?*, NBER Working Paper No. W15239, Cambridge, MA: NBER.
IPCC (2014), 'Climate-resilient pathways: Adaptation, mitigation, and sustainable development', in C.B. Field, V.R. Barros, D.J. Dokken, K.J. Mach, M.D. Mastrandrea, T.E. Bilir, M. Chatterjee, K.L. Ebi et al. (eds), *Climate Change 2014: Impacts, Adaptation, and Vulnerability, Part A: Global and Sectoral Aspects: Contribution of Working Group II to the Fifth Assessment Report of the Intergovernmental Panel on Climate Change*, Cambridge, UK: Cambridge University Press, pp. 1101–1131.
Jaeger, K.W. (2012), 'The double dividend debate', in J.E. Milne and M.S. Andersen (eds), *Handbook of Research on Environmental Taxation*, Cheltenham, UK and Northampton, MA, USA: Edward Elgar Publishing, pp. 213–229.
Kemfert, C., M. Kohlhaas, T. Truong and A. Protsenko (2005), *The Environmental and Economic Effects of European Emissions Trading*, DIW Discussion Paper 533, Berlin: DIW.
Kerkhof, A., H. Moll, E. Drissen and H. Wilting (2008), 'Taxation of multiple greenhouse gases and the effects on income distribution – a case study of the Netherlands', *Ecological Economics*, **67**, 318–326.
Klepper, G. and S. Peterson (2004), 'The EU Emissions Trading Scheme, allowance prices, trade flows, competitiveness effects', *European Environment*, **14**, 201–218.
Kuik, O. and M. Hofkes (2010), 'Border adjustment for European emissions trading: Competitiveness and carbon leakage', *Energy Policy*, **38** (4), 1741–1748.
McKibbin, W.J. and P. Wilcoxen (2008), 'The economic and environmental effects

of border tax adjustments for climate policy', Brookings Global Economy and Development Conference, Washington, DC.
Monjon, S. and P. Quirion (2011), 'Addressing leakage in the EU ETS: Border adjustment or output-based allocation?', *Ecological Economics*, **70**, 1957–1971.
Oberndorfer, U. and K. Rennings (2007), 'Costs and competitiveness effects of the European Union Emissions Trading Scheme', *European Environment*, **17**, 1–17.
OECD (2003), *Environmental Taxes and Competitiveness: An Overview of Issues, Policy Options and Research Needs*, Paris: OECD.
Öko-Institut (2013), *Support to the Commission for the Determination of the List of Sectors and Subsectors Deemed to Be Exposed to a Significant Risk of Carbon Leakage for Years 2015–2019*, Freiburg, Germany: Öko-Institut.
Paltsev, S. and P. Capros (2013), 'Cost concepts for climate change mitigation', *Climate Change Economics*, **4**, Suppl. 1.
Peterson, E.B., J. Schleich and J. Duscha (2011), 'Environmental and economic effects of the Copenhagen pledges and more ambitious emission reduction targets', *Energy Policy*, **39**, 3697–3708.
Peterson, S. and G. Klepper (2008), *The Competitiveness Effects of the EU Climate Policy*, Working Paper 1464, Kiel, Germany: Kiel Institute for the World Economy.
Ponssard, J.P. and N. Walker (2008), 'EU emissions trading and the cement sector: A spatial competition analysis', *Climate Policy*, **8** (5), 467–493.
Porter, M.E. and C. van der Linde (1995), 'Towards a new conception of the environment–competitiveness relationship', *Journal of Economic Perspectives*, **9**, 97–118.
Reinaud, J. (2008), 'Issues behind competitiveness and carbon leakage: Focus on heavy industry', OECD/IEA Information Paper.
Sartor, O. (2012), *Carbon Leakage in the Primary Aluminium Sector: What Evidence after 6½ Years of the EU ETS?*, WP No. 2012-12, Paris: CDC Climat Research.
Smale, R., M. Hartley, C. Hepburn, J. Ward and M. Grubb (2006), 'The impact of CO_2 emissions trading on firm profits and market prices', *Climate Policy*, **6**, 29–46.
Sue Wing, I. (2011), 'Computable general equilibrium models for the analysis of economy–environment interactions', in A.A. Batabyal and P. Nijkamp (eds), *Research Tools in Natural Resource and Environmental Economics*, Singapore: World Scientific Publishing, pp. 255–306.
UNFCCC (2015), Adoption of the Paris Agreement, UNFCCC Conference of the Parties, Twenty-first session, Paris, 30 November – 11 December.

17. Claim the sky!

Robert Costanza

The atmosphere is a community asset that belongs to all of us. The problem is that it is currently an open access resource – anyone can emit carbon dioxide into the atmosphere with no consequences to themselves – but huge cumulative consequences to the climate and the global community. Many agree that charging companies and individuals for the damages their emissions cause, for example a comprehensive carbon tax or cap/auction/dividend/trade system, would encourage companies to cut emissions. However, despite some interesting regional experiments, implementing this kind of system via international negotiations at the global scale has proven close to impossible.

Global civil society can change this if it *claims property rights* over the atmosphere. By asserting that all of us collectively own the sky we can begin to use the legal institutions surrounding property rights to protect our collective property, charge for damages to the asset and provide rewards for improving the asset. The public trust doctrine is a powerful legal principle that supports this argument. In her book, *Nature's Trust: Environmental Law for a New Ecological Age*, legal scholar Mary Wood describes how the public trust doctrine has its roots in ancient Roman law and occurs in the many legal systems derived from it. The doctrine holds that certain natural resources are to be held in trust as assets to serve the public good. It is the government's responsibility as trustee to protect these assets from harm and maintain them for the public's use. Under this doctrine, the government cannot give away or sell off these public assets to private parties. The public trust doctrine has been used in many countries in the past to protect water bodies, shorelines, fresh water, wildlife and other resources (Sax, 1970). However, as Mary Wood forcefully argues, the time has come to expand the purview of the doctrine to cover all of the critical natural capital and ecosystem services that support human well-being, including the atmosphere, the oceans, and biodiversity (Wood, 2013). She argues that governments have been shirking this responsibility to protect "nature's trust" and need to be required to fulfill their duty, including claiming damages. This problem is complicated by the fact that the atmosphere is a global asset, shared by and benefiting all the world's sovereign nations. However, from the nature's trust perspective, the fact that all the world's nations are *co-trustees* makes them no less responsible for protecting the asset than individual sovereigns are for protecting assets like shorelines or open water bodies that occur exclusively within their borders.

The nature's trust idea significantly changes the whole discussion about how to deal with climate disruption. Rather than national governments negotiating with each other about emissions reductions, governments should see themselves as co-trustees with a fiduciary responsibility to protect the atmospheric trust. To do this they can claim damages from the private interests that are causing the problem. As Wood notes: "Trustees have an affirmative obligation to recoup monetary damages against third parties that harm or destroy trust assets" (p. 185). If an oil spill occurred in the oceans, governments would collect natural resource damages for cleaning up the mess. Yet they sit idle in face of a catastrophic "spill" of carbon dioxide into the atmosphere.

Holding climate polluters accountable for their damage is more straightforward than it might seem. Approximately 90 companies globally are responsible for introducing two-thirds of the carbon emitted into the atmosphere (http://www.theguardian.com/environment/2013/nov/20/90-companies-man-made-global-warming-emissions-climate-change). This means that damage claims could target a relatively small number of private interests. In addition, all governments would not need to agree in order to employ this tactic. Since all governments are co-trustees in the global atmospheric asset, a subset of nations could bring the damage claims. Damages achieved from these legal actions could fund restoration projects in those same countries, provided that they are certified to draw down atmospheric carbon or expedite the transition to non-nuclear, renewable energy. In addition, governments could charge for ongoing damages via a carbon tax or other mechanisms.

But given that governments have not acted on their own, well-planned civil society pressure will be required to support governments to act, and to counteract the inevitable corporate resistance. A concerted effort to "claim the sky" as a public trust on behalf of all of global society, in combination with the solid legal framework provided by Wood's work on the public trust doctrine, may just do the trick. The civil rights movement in the U.S. was based on equally solid legal principles, but required coordinated social activism to achieve success.

Let's establish a broad coalition of individuals and groups and publicly claim on the web and in other venues that the atmosphere belongs to all of us, and our descendants, collectively and that we demand that the polluters pay for recovering and maintaining our atmosphere. We are going to press our beneficial rights through our governments and against the corporations themselves. We will establish an Earth Atmospheric Trust to qualify restoration projects (projects that achieve carbon drawdown through soil sequestration and reforestation, or that promote transition to a renewable energy infrastructure). These are not "carbon offset projects." We have to go beyond offsets to demand *cleanup* of the atmosphere (Moomaw, 2014). The Trust will maintain a financial accounting and carbon accounting of projects funded by corporate polluters to carry out their liability to the citizen beneficiaries. A public outreach campaign to build pressure will start by sending invoices to the polluters themselves for past and ongoing damages (see the example in Appendix 17.1).

Imagine the following:

- An Earth Atmospheric Trust is established to collect past and ongoing damages and to certify and track restoration projects funded by corporations who are liable for carbon emissions, and to directly fund restoration projects.
- Groups of school children deliver invoices to the corporate headquarters of major fossil fuel companies with major media coverage.
- 350.org and other groups take up the campaign to finally get movement toward their goal.
- The occupy movement reinvigorates itself around this campaign.
- Major international NGOs like Conservation International, the World Resources Institute, lend their support to the campaign.
- Major media personalities, like Robert Redford, Harrison Ford, Brad Pitt, Leonardo DiCaprio, Cameron Diaz, and others (http://www.examiner.com/article/42-celebrities-who-care-about-the-environment), join the campaign.
- Michael Moore makes a documentary about the campaign.
- . . .

Nature's trust has the potential to be a significant new legal and social force in the battle to bring human activities on the planet into compliance with "nature's laws." It is far past time for the people of planet Earth to claim the future by claiming the sky.

BIBLIOGRAPHY

Barnes, P. (2006), *Capitalism 3.0: A Guide to Reclaiming the Commons*, San Francisco, CA: Berrett-Koehler.

Barnes, P., R. Costanza, P. Hawken, D. Orr, E. Ostrom, A. Umana and O. Young (2008), 'Creating an Earth Atmospheric Trust', *Science*, **319**, 724.

Moomaw, W. (2014), 'From failure to success: Reframing the Climate Treaty', *Fletcher Forum of World Affairs*, http://www.fletcherforum.org/2014/02/10/moomaw/.

Sax, J.L. (1970), 'The public trust doctrine in natural resource law: Effective judicial intervention', *Michigan Law Review*, **68**, 471–566.

Wood, M.C. (2013), *Nature's Trust: Environmental Law for a New Ecological Age*, Cambridge, UK: Cambridge University Press.

APPENDIX 17.1

Date: *Soon*

INVOICE

To: *Company XX*

Under the public trust doctrine, you are hereby assessed for damages to the Global Atmospheric Commons

$125 per cumulative ton of CO_2 equivalent introduced into the global economy from *date* to *date* (based on the latest IPCC estimates of damages), plus ongoing damages.

= $xxx,xxx,xxx,xxx.00

Deposit this amount in the Global Atmospheric Trust Fund within 90 days or face legal action and sanctions by the shareholders of the Trust – the people of Earth.

Monies in this fund will be used exclusively to maintain and improve the Atmosphere for the benefit of all shareholders, present and future. These uses include, but are not limited to:

1. Investment in community owned, low-carbon emitting renewable energy sources such as wind and solar
2. Investment in carbon sequestration projects including in forests, soils, and wetlands
3. Investment in urban infrastructure improvements to reduce car use and improve building energy performance
4. Investment in technology development to enhance and accelerate the above.

(As an alternative to paying this invoice in full, you may invest an equivalent amount in projects that have been approved and monitored by the Trust in one of the categories above).

Signed

Shareholders in the Earth Atmospheric Trust

Index